TABLE OF CONTENTS

Project Management for Data Conversions and DATA MIGRATIONS

A Data Conversion Methodology and Guide to Converting Data for Mission Critical Applications

by

Charles R. Scott

authorHOUSE™
1663 Liberty Drive, Suite 200
Bloomington, Indiana 47403
(800) 839-8640
www.AuthorHouse.com

This book is a work of non-fiction. Names of people and places have been changed to protect their privacy.

First published by AuthorHouse 10/18/04

ISBN: 1-4184-5216-5 (sc)

Library of Congress Control Number: 2004096044

Printed in the United States of America
Bloomington, Indiana

This book is printed on acid-free paper.

“THE POWER OF THE MIGHTY IS BALANCE”

C. R. Scott

PROJECT MANAGEMENT FOR DATA CONVERSIONS AND DATA MIGRATIONS

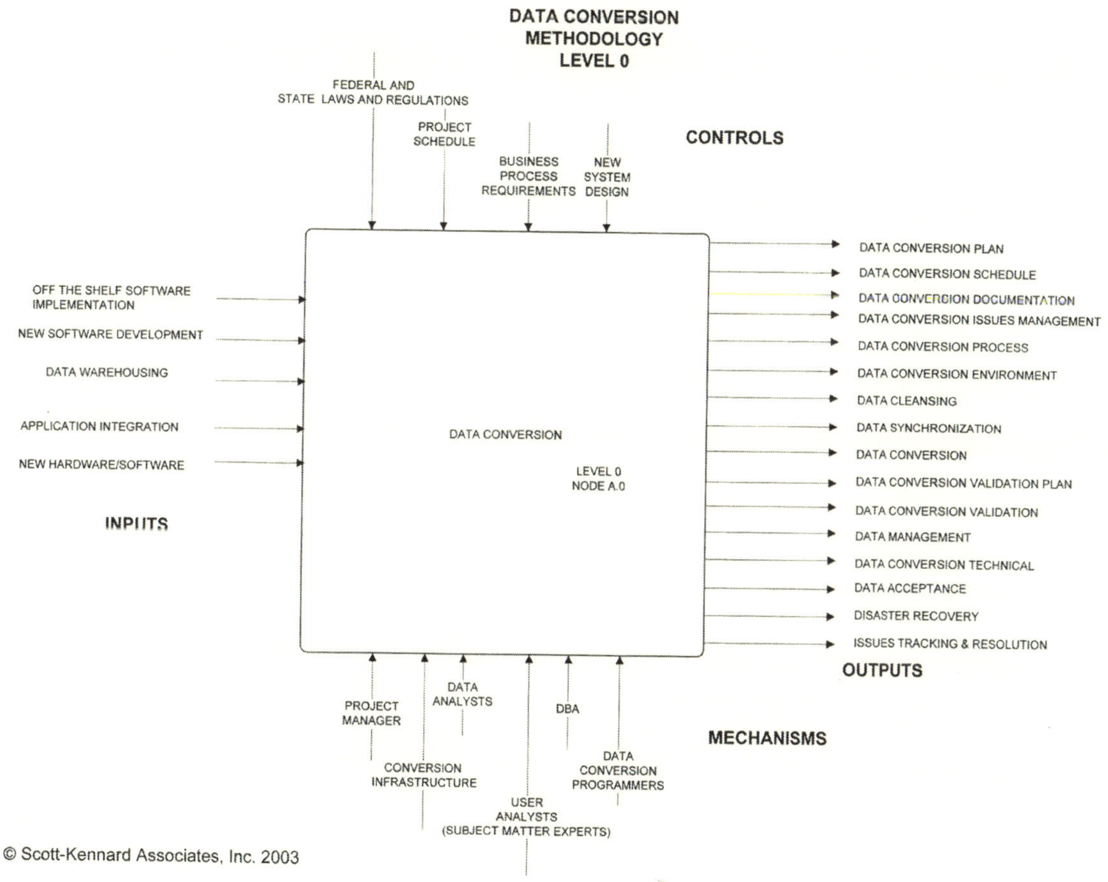

A DATA CONVERSION METHODOLOGY AND GUIDE TO CONVERTING DATA FOR MISSION CRITICAL APPLICATIONS

FORWARD

Data Conversion can mean many things to many people. If one were to search the internet for the key words "data conversion", one would find that there are about six million (6,000,000) entries located for the search. These entries fall into the following categories:

- Tape, disk, CD's, Zip, Jag,, paper, et. conversions
- Data duplication
- Optical disk conversion
- Mail order mapping
- Data recovery
- Programming language conversion
- EBCDIC to ASCII and vise versa
- Data mapping
- Data reporting
- CAD data conversion
- Analog to digital data conversion and vise versa
- Scanning and digitizing
- Currency conversion
- PDF,RTF,XML,HTML, SGML, etc. conversions
- Software to assist in conversion

Included in the list one will also find a host of variations of the above plus other data conversion topics not mentioned here. This makes for a very confusion mix when one is about to embark on a major project to convert the legacy data within an entity to a new application system. To a serious observer, there does not seem to be a fit.

Another round of confusion arises when data conversion tasks are embedded within, and co-mingled with the new application implementation projects task, never getting the attention they deserve. Application implementation Project Managers should realized that “nothing happens until the data is converted” and converted accurately. Recognition of this fact would generate more focus on the data conversion tasks within the project. An added burden in getting the data converted may occur if as sometimes happens, leadership spots are assigned politically and not by qualifications. Lack of experience and lack of listening skills will lead to increased cost, an increase in implementation times, and a decrease in user acceptance.

Finding internal resources in an IT department to participate in a data conversion project is not always the easiest thing to do. If someone in the past has participated in a data conversion project, they usually do not want to do another simply because of the difficult work and long hours entailed in a conversion. They may also have been promoted since the last conversion project in which they participated and cannot take the time from their daily tasks to participate.

This methodology concerns itself with addressing all aspects of data conversion. It provides a methodology for project managers to assure the least amount of risk in performing data conversion for a new application implementation project, or for a data migration project, and accomplished it successfully with the least amount of stress, cost, and time.

This book is necessary because of horror stories that crop up in many data conversion projects. Consider an experience that happened at a Fortune 500 company: After conversion weekend, the receiving dock had trucks backed up about a mile long by the following midweek because the purchase orders could not be found on the converted data to make the receipts.

This book then, provides a methodology for those projects that are embarked on data conversion for mission critical applications, like ERP (Enterprise Resource Planning), EAP (Enterprise Asset Planning), data migrations (from one software release to the next), and data warehousing. This implies that this methodology, or parts of it, will work for any data conversion project.

It is assumed that the user of this book will have some knowledge of Project Management, either through on the job training or through formal training. Evaluating the schedule, adding tasks, and adding resources on a frequent basis is a key component of this methodology. For the more informed, it is wise to let the data conversion lead manage the dollars for the data conversion part of the project.

This methodology contains years of experience and trial and error attempts at data conversion. It is not intended to be foolproof or provide all the answers for all situations. However, it is a methodology that will work for you on the data conversion/ data migration project.

Since time is a valuable commodity to each of us, the 400 pages of personal stories and pontifications have been omitted from this document. Contained herein are only the net statements which will contribute to a successful project.

INTRODUCTION

Data conversion is never easy. It is often under budgeted, under staffed, and behind schedule. Usually when data conversion is mentioned as part of a project, the first thing that gets discussed is data mapping, when data mapping is only one of the components of data conversion.

The Data Conversion Methodology that is provided within this document considers all the components of data conversion. It has been brewing since computers came into existence, and tested in several data conversion environments and it works! It is possible that the project will not need all the components of this methodology, or it may be that you have components in the project that are only partially covered in this methodology. Either case is addressed by this methodology, and if you find it provides exactly what you need, so much the better. The intent here is to provide a base from which to generate a data conversion plan and schedule for successful execution, regardless of the project.

Additionally, the diagrams are not true IDEF0[1] diagrams, in that each node does not show all the inputs, controls, outputs, or mechanisms. To represent them in this fashion would make the diagrams very confusing to read. Instead, the diagrams represent tasks within a project from which the user can produce an IDEF0 diagram if so desired. In fact, to make the scheduling part of the project easier, and to create a realistic project environment, each diagram should be expanded to include all aspects of the IDEF0 in the planning phase.

The Level as shown in each diagram indicates the level in a project schedule. The node indicates sub levels within the level.

SOME OVERLOOKED DATA CONVERSIONS CONSIDERATIONS

1. **Pay me now or pay me later.**

   In many projects, in an effort to reduce the cost of the overall implementation, shortcuts are taken, or “right before your face” new found tasks are skipped in order to meet the schedule. A prime example of this is data clean-up. It is very obvious at times that some of the data being converted needs to be cleaned up prior to conversion or during conversion so that when the system becomes operational after the conversion weekend, it is usable. If the time and effort is not made to clean up the data before or during conversion, then the data will have to be cleaned after conversion or the system will be unusable in many ways. When the data to be cleaned is a planned task after the conversion weekend is over, then there must be a planned task added to the schedule. The planned cleanup task must be taken as seriously as the initial conversion effort.

   In reality, skipped task that are critical to the operation of the system will have to be paid for either before conversion weekend, or after conversion weekend. These costs should be planned for as much as possible at the beginning of the project by the project planning team.

2. **Resources.**

   For the data conversion project, just having programmers available is not enough. The process, data analyst, application analyst, data base administrators, and data testers are all required for a successful conversion. There are too many cases where the resources allocated to a data conversion project are other left over project resources, or they are the least busy, or they are about to get laid off for one reason or another. The resources should be among the best available and have gone through a data conversion project before. If not, it is a good idea to have expert outside resources to help create, manage and execute the data conversion plan. The outside expert resources probably will come from the software vendor, or a vendor that specialized in data conversion or Project Management activities. These resources, though expensive, can actually shorten the project and solve many of the problems that will arise during the conversion process, thereby resulting in less cost for the project.[2]

3. **Managing the team**

   The nature of the data conversion teams tasks, and the mix of technical skills required to complete the tasks, requires that the manager of the data conversion team be open, flexible, and not in a CYA mode. The truth of the matter is that there will be mistakes made during the conversion process and pointing fingers will do no good in resolving the problems. What must be done is to identify the problem, document the problem, and resolved it. Therefore, communications skills are a must. The team must understand the schedule at all times and each member of the team must know where they fit in the giant puzzle.

4. **Costs and Time**

   If you use the included methodology or some portion, and if you plan for all the tasks associated with the diagrams, it is a good idea to then double the cost of the data conversion project and increase the time by one half. This is required because of all the unknowns associated with the data conversion. You have the users, analyst, programmers, and sometimes data that has never been understood, hardware resources that are not 100% available and configuration changes which contribute to the added cost and time. However, if you have not used the included methodology and have derived the schedule based on the last project, then it is a good idea for you to triple the first cost estimate of the data conversion project and double the time.

NOTE: This formula is not cast in concrete, but it will produce a data conversion cost and time that is generally more realistic to use as a planning tool.

THE

METHODOLOGY

Chapter 1

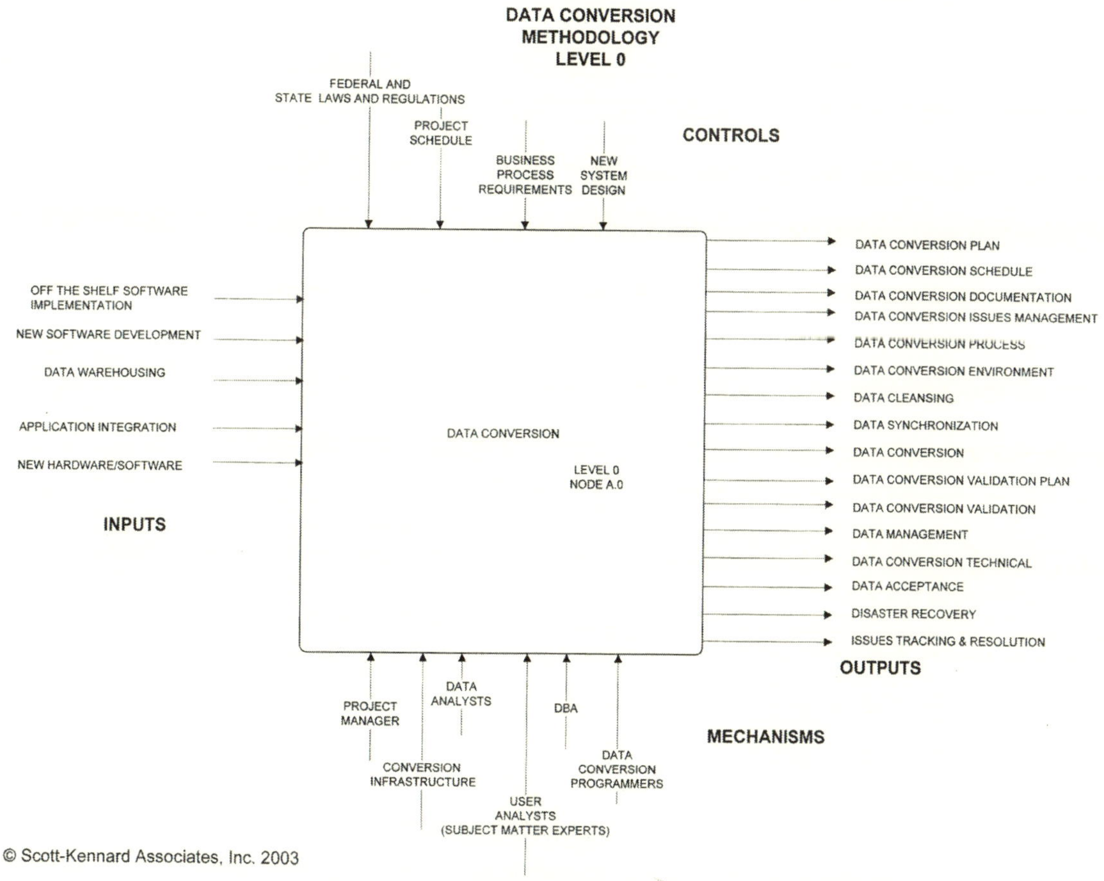

LEVEL 0, NODE A-0 Data Conversion

INPUT

Before starting a plan for data conversion, one should know the reason that they are doing data conversion, or the requirement that generates the data conversion. Some requirements that drive data conversion are shown in the Level 0 diagram above as inputs. The requirements statement should be very specific and delve deeper into the inputs which will shed light on the enormity of the tasks ahead.

State the requirements for the data conversion project here.

CONTROLS

Within any data conversion project, external forces must be considered which will influence the data conversion plan. It is important to note early in the plan what these forces are and include them as part of the data conversion plan and schedule. These are required forces. If there has been a Business Process Redesign, then the output of this process must be considered when producing the plan. Any of the Control information needs to be documented in this section of the plan.

List all the controls that will influence the data conversion project. Use the A-0 node as a guide.

OUTPUTS

The output of the conversion project should obviously be the implemented data. To get through the maize, a series of plans need to be in place. These plans may be as extensive as a new version of the dictionary, or they may be as simple as a paragraph. You must choose for the plan. However, each plan should have a delivery date and will comprise the steps in the project tasks schedule. Each output should be considered a deliverable and scheduled as such.

List all outputs expected for the data conversion project. Use the A-0 node as a guide.

MECHANISMS

What resources do you need to get through data conversion? These resources should be attached to the task in the project schedule. However, identify as many as possible at the beginning of the project, even if you overstate the resource requirements, so that the scarce resources can be claimed and obtained.

List all resources expected for the data conversion project. Use the A-0 node as a guide.

When you have completed this section, you should have a modified A-0 node that drives the rest of the planning for the project.

PROJECT NOTES

Chapter 2

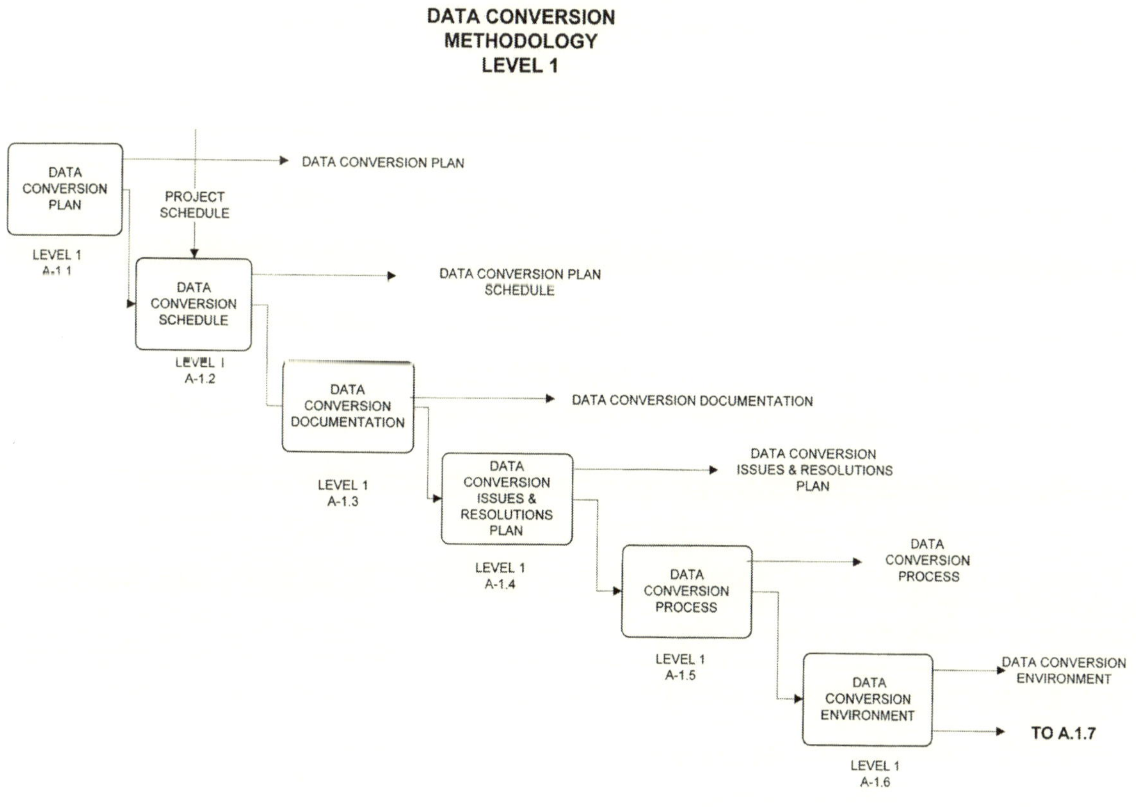
DATA CONVERSION
METHODOLOGY
LEVEL 1
DATA CONVERSION PLAN
LEVEL 1
A-1 1
DATA CONVERSION PLAN
PROJECT SCHEDULE
DATA CONVERSION SCHEDULE
LEVEL 1
A-1.2
DATA CONVERSION PLAN SCHEDULE
DATA CONVERSION DOCUMENTATION
LEVEL 1
A-1.3
DATA CONVERSION DOCUMENTATION
DATA CONVERSION ISSUES & RESOLUTIONS PLAN
LEVEL 1
A-1.4
DATA CONVERSION ISSUES & RESOLUTIONS PLAN
DATA CONVERSION PROCESS
LEVEL 1
A-1.5
DATA CONVERSION PROCESS
DATA CONVERSION ENVIRONMENT
LEVEL 1
A-1.6
DATA CONVERSION ENVIRONMENT
TO A.1.7

DATA CONVERSION METHODOLOGY
LEVEL 1
(Continued)

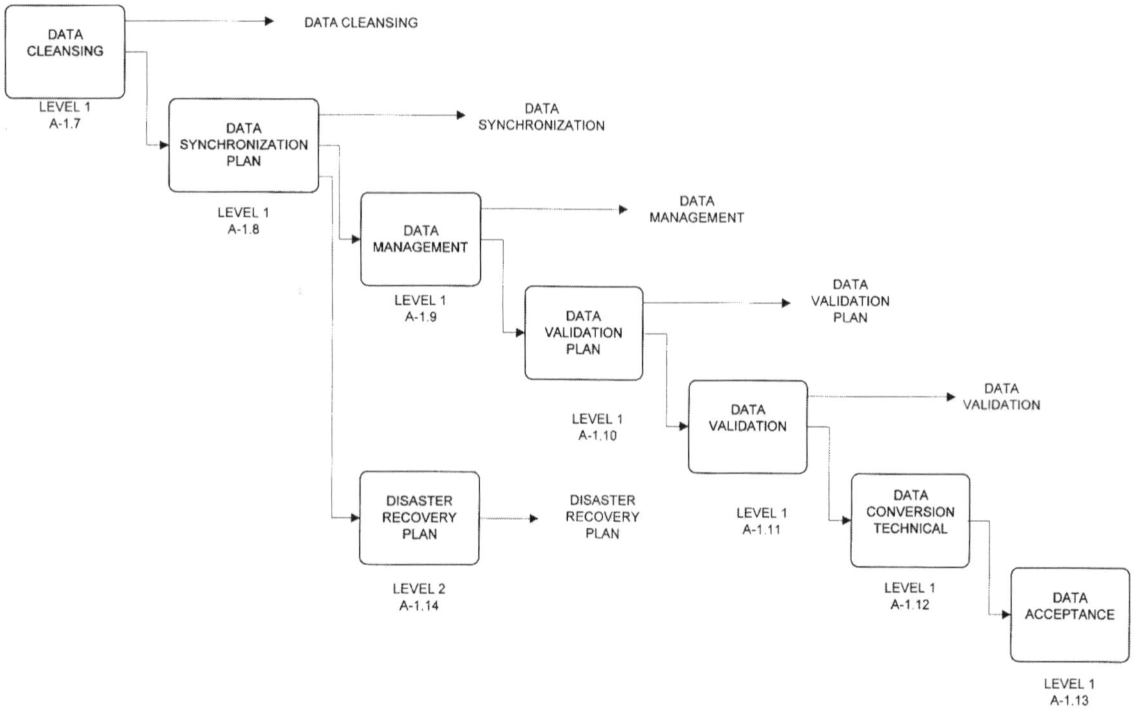

DIAGRAM 2 LEVEL 1 - Methodology

LEVEL 1 Data Conversion Plan

The startup planning that **must** be done at the beginning of each data conversion project will make life easier for everyone. The preliminary and startup work associated with the planning, will represents the basis for the project. If the plan is not put into place, and questions and issues are not resolved up front, many hours will be wasted on the project. You will find that there will be duplication of effort, meetings called to resolve the same issue over and over again, and each group going its own way to perform a task that has no consistency to obtaining satisfactory results which adds havoc to the project.

Tasks for each of the areas represented in the Methodology diagram are given in detail in the following explanations.

PROJECT NOTES

Chapter 3

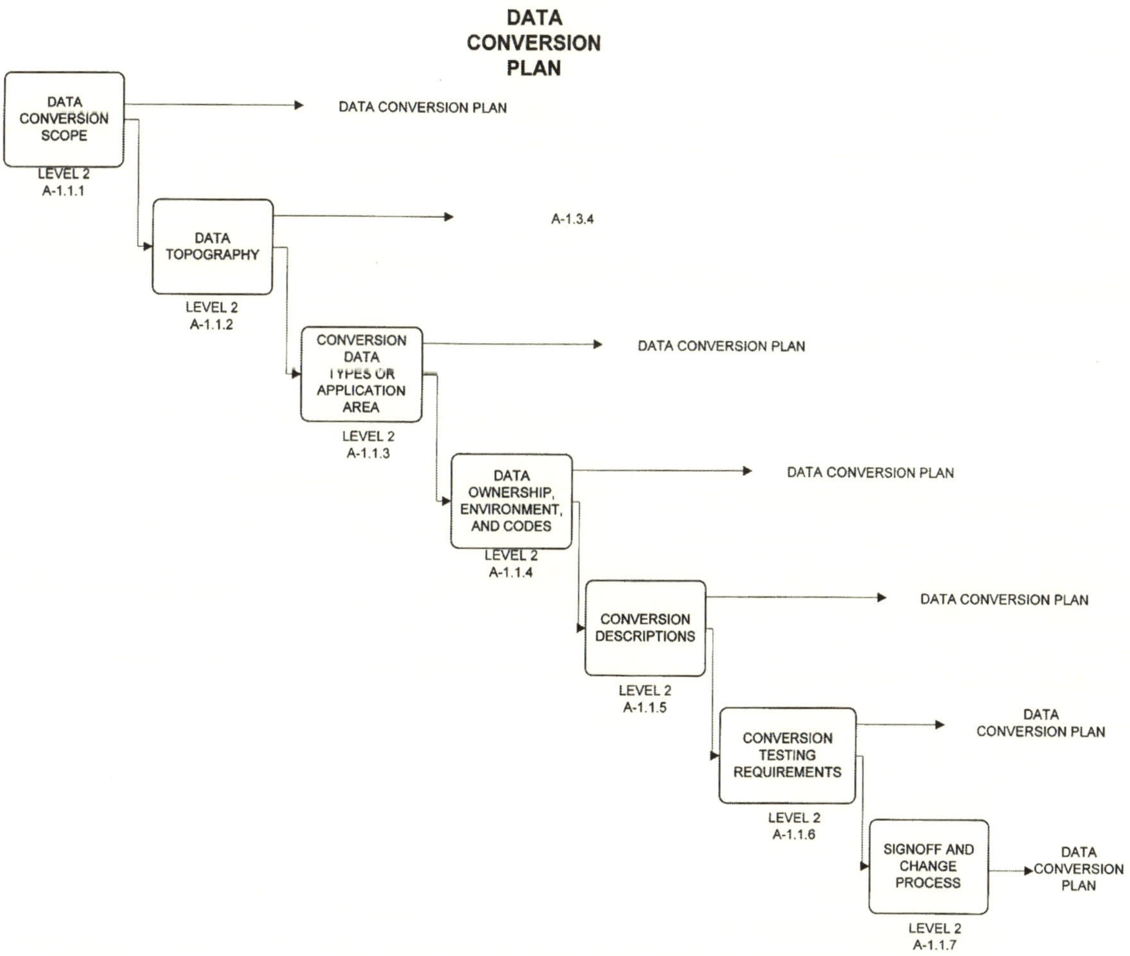

DIAGRAM 2 LEVEL 2 - Data Conversion Plan

PURPOSE:

The purpose of this section is to provide a plan for the data conversion team. Completing this section will establish the ground rules for the conversion project and can save many project hours.

LEVEL 2, NODE A-1.1.1 DATA CONVERSION SCOPE

Based upon the Input to the data conversion project, a scope must be defined as part of the project plan. The scope will list all the data required to meet the objectives of the input to the project. If it is work orders for a specific division of the company, then only that division's data will be listed in the scope. If more than one division of a company's data of a specific type is to be converted, and the data lies in two separate locations on the current computer environment, then there must be a separate entry in the scope document for each. In some cases, when writing the plan, it may not be possible to determine the exact scope, because further evaluation of the data is needed. If this is the case, it should be so stated in the plan and all of the possible data listed as candidates for conversion. In addition, activities must be identified in the schedule for the evaluation of the data so that the scope can be finalized prior to the beginning of the data mapping exercise.

LEVEL 2, NODE A-1.1.2 DATA TOPOGRAPHY

Topography is a description of all the surface features, natural and artificial, of a particular region. It is the science of drawing maps and/or diagrams which represent these features.

For any data conversion project, it is important to describe the features of the data. To adequately describe these features, one must take a global view of all the data that has to be converted, transferred, or moved. For some conversions, this could be simple and does not require a lot of thought or documentation. For instance, if we were going to convert text data on a PC to a database on the same PC, we would probably have a good idea what the text data looked like, and what the data base table is to be. In addition, we would probably use some import tool to move the data from the text file to data tables. We could use a simple diagram as shown below to represent the move.

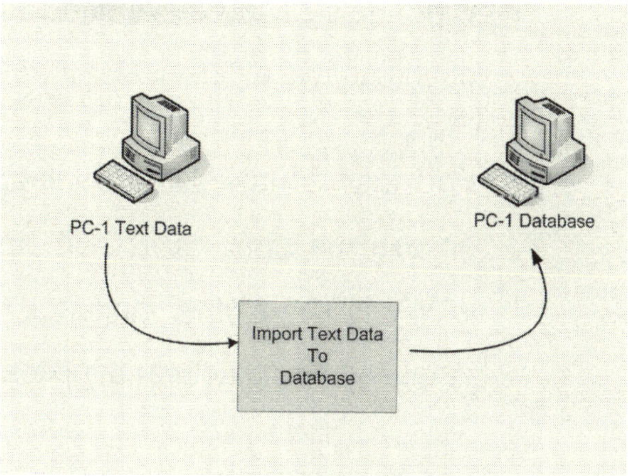

By providing this diagram (Topography), as simple as it may appear, several things are obvious:

1. The data to be converted is on PC-1, a personal computer or workstation
2. It is located on disc on PC-1.
3. The conversion will be from a text file to database.
4. No special communication protocol is required.
5. No special security is needed.

Of course, most conversion projects are not this simple. The level of complexity gets higher with each additional hardware component, database component, or communication protocol. In documenting the topography, the following requirements must be identified and documented:

Location of data. Where does the data reside that must be converted? Is it on a mainframe computer, on a LAN server locally, in another state, in another country, does it reside on a desktop, a piece of paper, or is it some combination of all of the above?

Hardware Platform/Medium What type of computer, disk, tape, or medium does the data resides on. There may be a big difference in what the data looks like on two different tape or disc drives on two different Mainframes.

Software Platform For each source, list the language that is used to access the data. Is it COBOL, FORTRAN, Basic, JAVA, or some other language? Is it a utility that the database manufacturer supplies?

Data Base Platform For each data source, list the database or data type, which drives the access of the data. Is it Relational, Hierarchical, flat, or some other format?

Communication Platform For each source, list the communication vehicle for moving the data. How are you going to get the data from the source to the location where it has to be transformed or loaded? How fast is the communication platform? Does the line speed have to be taken into account for the conversion process?

Security Can the data be accessed remotely or locally when needed? Any special access required?

Skills List any special skills required to access, transform, move, or store the data at the source. Beware of Hugo. Hugo is the "little ole man" sitting in the back room and hasn't talked to anyone in the past five years.

However, Hugo developed this **critical** application 10 years ago that he maintains and only he knows anything about it. If you are going to need Hugo, be sure to include him early in the project.

Topography, as applied to data conversion, is the beginning of understanding of the tasks that lie ahead for the project. Without it, schedule delay and cost over run for the project is guaranteed.

A typical topography is shown **below**. A description for each heading above must be made for the graph. Add as much verbiage as needed to define the layout of the data. If a user has a book or piece of paper or some other manual process that is used for recording data, include that in the topography. It is a good idea to obtain sign-off from the user representative after the topography is completed. This topography will provide input to:

- Scope of data to be converted
- Skills required
- Conversion process
- Timing of moving data
- Security requirements
- Software utilities
- Personnel required at each site

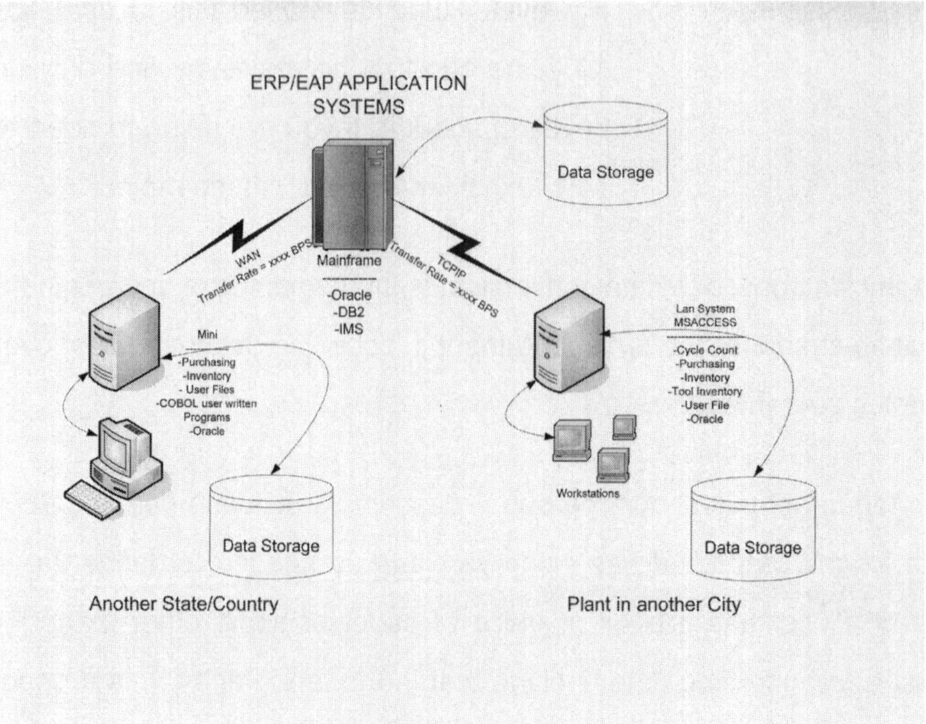

This Node feeds into NodeA-1.3.4

LEVEL 2, NODE A-1.1.3 DATA TYPES OR APPLICATION AREA

When embarked on a data conversion project, the conversion process does not take place all at the same time for all the application areas. It is therefore required that output from the Topography of the data be fed into the different applications areas so that conversion can be planned in a coordinated effort. Somehow, it is not right to get the data for an Inventory application from area A without planning on retrieving the Purchasing data which may or may not be in area A, but may be located in area B and coordinating the two.

The application area can be described as one type of data type. Another description of data type relates to the data element, whether it is numeric, character, how many decimal places, etc. For purposes of the data conversion plan, the data types refer to the application area(s). The data types for the data elements are described in the data mapping document.

LEVEL 2, NODE A-1.1.4 DATA OWNERSHIP

There is no intent to be ridiculous by citing the obvious, but this task is often overlooked or assumed, when in reality, having one take responsibility for legacy or converted data is difficult. It is obvious that there are systems that have evolved over the years and the question has never been addressed as to who owns the data. Then there are those systems where it may not be obvious who owns the data. In any case, get ownership of the Legacy data to be converted assigned because decisions have to be made regarding archiving, legacy shutdown, and acceptance signoffs on the new system.

One thing that has complicated the data ownership issue is the integration of applications that has taken place over the past 20 years. There are elements of data within each application area where the owner clearly is outside the application area, but their data is critical to the application area. It is therefore recommended that a chart be constructed during the planning phase that identifies those individuals by name. A sample of this chart could look as the following:

APPLICATION AREA	DATA OWNER	DATA ELEMENT OWNER
Purchasing	Purchasing Manager	
- Vendor		Accounting Manager Procurement Engineering
- Part or Item Number		Engineering
- Required Due Date		Production Manager (Based on Schedule)

The example is the Purchasing application area. The data manager in this case is the Purchasing manager. For conversion purposes, the purchasing manager is the responsible contact. However, there are some data elements within purchasing in which the purchasing manager has no control because the Purchasing Department is a user of the element. These data elements must be identified as in the above example where the Vendor is used by the purchasing department but is set up and maintained by the Accounting Department. Furthermore, it is possible that quality constraints are placed on the part or item and Procurement Engineering has to decide if that vendor is eligible as a vendor. The Purchasing Department will have input to the Accounting Department or Procurement Engineering Department, but are not responsible for the data elements.

A chart as shown above for the application area being converted is mandatory for the data conversion team. It will reduce issue resolution time, and add consistency to the conversion effort.

LEVEL 2, NODE A-1.1.5 DATA CONVERSION DESCRIPTIONS

The data conversion descriptions explain how the data will be moved from platform to platform, system to system, or network to network. It will describe data extract programs and tools to be used for manipulating the data on the appropriate platform for which the data exist. If special considerations are required because of the location of the data, then these are described. The purpose of this node is to establish a standard for moving the data and making sure the tools are provided and the data infrastructure is planned and established based on the topography of the process.

This is especially important where there is more than one technical resource involved. Too often, the programmer will have their own method of:

a) Moving the data from one location to another
b) Transforming data from one type to another
c) Looking at data in its source form
d) Looking at data in its target form
e) Justifying the accuracy of the data

An analysis of the available tools[3] in the client environment, and an analysis of which tools are required to perform the conversion will lead to the descriptions of the data conversion. It is required that once the descriptions are completed, all technical resources will adhere to the standards set forth. If there are descriptions that a resource is not comfortable with and feel that their individual method is better, then that resource should either be trained to use the new descriptions, or assigned to another task in which they feel comfortable.

This is important, because as a manager of the conversion team, it is vital to be able to trouble shoot data problems that may arise. If there is consistency in how each technical resource will perform their process, then solving a problem for one, solves it for all. However, if each has their individual personalized way, then problem solving becomes longer, increasing the task of the conversion.

LEVEL 2, NODE A-1.1.6 DATA CONVERSION TESTING REQUIREMENTS

The data conversion testing requirements must be stated as part of the Data Conversion Plan. Some requirements to consider are:

- Location (Machine or Region) in which testing is to be performed in the various stages of

- Unit testing. Testing of the programs one at a time to make sure they perform according to the specifications, including data transformations and target

- System Test. Conducting test where all the units are executed in sequence and/or in tandem. The system test usually is conducted against a user written script that mimics a business process and includes all inputs and outputs in the process.

- Performance Testing. Testing to make sure the system will perform according to re-established criteria. The most common performance test is for response time.

- Test Scripts Test scripts should be generated for each test scenario.
- Test Data Test data must be provided. The test data may be a version of the converted data, even if it is not in its final form. However, the buy-off version of the testing should be performed with the final conversion data.

LEVEL 2, NODE A-1.1.7 SIGNOFF AND CHANGE PROCESS

Within any project, the sign off process can get very lengthy. Once the sign off process is complete, if changes are required, then the process gets more complicated. There should be a simple process defined at the beginning of the data conversion project that is designed to get signoffs and approvals as expeditiously as possible.

What generally happens is that no signoff process or approval mechanism is defined in the plan and the data conversion manager ends up making decisions that may not be accurate, but to keep the project moving forward, makes a data judgment that will most likely cause extra work in someone's schedule later in the project.

Chapter 4

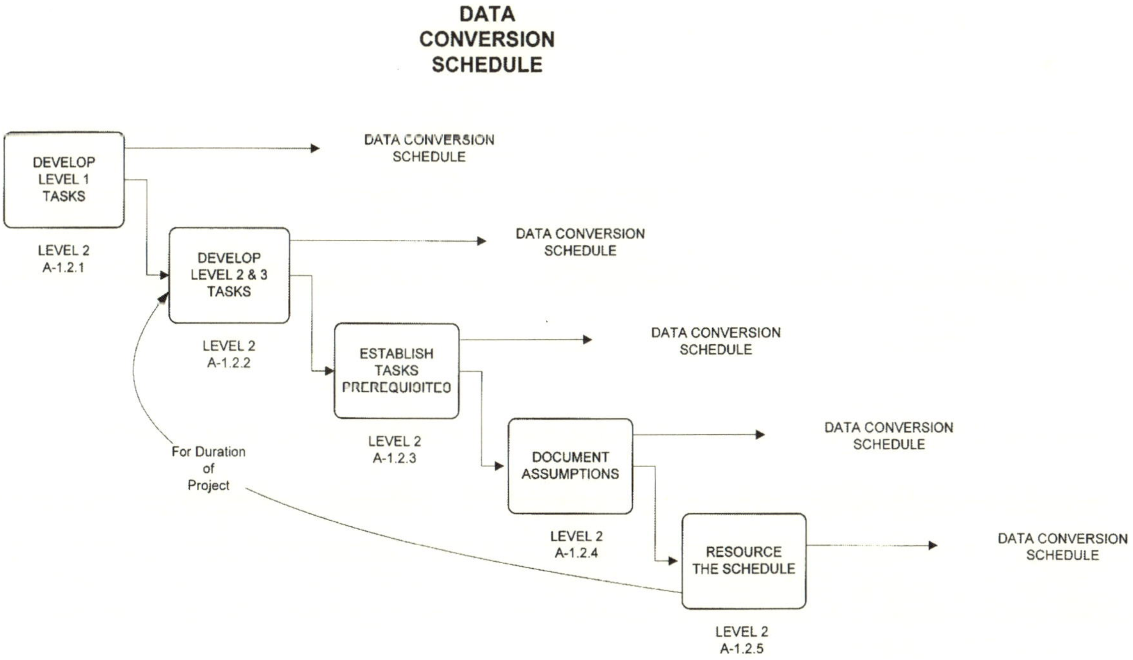

DIAGRAM 3 LEVEL 2 - Data Conversion Schedule

PURPOSE:

The purpose of this section is to provide a schedule for data conversion. This schedule may be part of the overall Project schedule, but must be managed by the data conversion manager. In addition, included in the plan and schedule are all tasks that you can think of where others are depending on the data conversion project to complete, so that they can perform their tasks (Like training, or testing). Also have task up front in the schedule where you need task performed by others to complete the data conversion (Like region management or region tuning).

It has been my experience that having control of the data conversion project plan makes reporting easier. As with all projects, especially data conversion projects, new task must be added, usually on a daily basis, because of the changing nature of the requirements. Or, if the initial plan is not thoroughly reviewed for missing elements or accuracy, daily maintenance of the schedule must be performed.

It has also been my experience to use the data conversion schedule as a way to keep track of what has to be done, what has been done, who is waiting on data conversion task to complete, and who needs to be contacted to get other task accomplished to meet the schedule. If the times per task are not accurate prior to performing the task, the actual times must be updated as the tasks are completed which makes tasks cost possible. The only problem reporting time against a task is that estimating percent complete is often incomprehensible. Programming task tend to stay at 90% complete, and hours pour on the task for weeks without increase in progress. How many times have you heard the "I am complete except for" statement made and how many time does the "except for" turn out to be another 90% time spent on the task?

LEVEL 2, NODE A-1.2.1 DEVELOP LEVEL 1 TASKS

Level 1 task for the data conversion schedule can be taken from this methodology. Each level one task represents a deliverable to the project and represents a high level activity of what has to be done to complete the data conversion.

LEVEL 2, NODE A-1.2.2 DEVELOP LEVEL 2, 3 TASKS

All tasks below level 1 represents the steps required to accomplish the level 1 task. By using this methodology as a reference, many of the tasks can be identified for the schedule. It is up to the data conversion project manager to refine the tasks and schedule to meet the data conversion requirements.

LEVEL 2, NODE A-1.2.3 ESTABLISH TASK PREREQUISITES

As with any project plan or schedule, some task must be completed before other tasks are started or completed, while other tasks are in progress, or at the same time as another tasks. Make sure you have reviewed the tasks to establish these relationships.

There are also tasks outside the data conversion schedule that depend on the data conversion task being completed by a specific date. For instance, data may be needed for training, or to meet a systems requirement. These should be documented in the schedule.
Last but not least are those tasks that must be completed but are outside the data conversion schedule before tasks within the data conversion schedule can be started. Setting up workspace for the data converters, obtaining necessary equipment, and getting the necessary systems work performed are examples.

Just make sure that these interdependent tasks are identified and communicated. It is very embarrassing, as part of the conversion team to have a scheduled task about to begin, and some other sub-project (like test data creation) not started and throw the project off schedule.

LEVEL 2, NODE A-1.2.4 DOCUMENT ASSUMPTIONS

Assumptions that are made in relationship to the data conversion project and to each task should be documented as part of the task notes. A common assumption we make is that work will be done by the people resources working 8 hour days, five days a week, and that all required people resources will be available. These assumptions may be true for some projects, but for many other projects, this will never happen. Therefore, these assumptions should be documented so that as the project progresses, these assumptions can be altered, or the schedule adjusted based on reality.

LEVEL 2, NODE A-1.2.5 RESOURCE THE SCHEDULE

For each task, use the "best judgment" to determine which resource, as well as how much of the resource you will need to complete the task. "Best judgment" is recommended because no two data conversion projects are the same. Resource skills vary and the amount of work required to perform a task varies. It helps if you have managed the task before, but getting exact numbers for "how much" and "who" is not always easy.

Chapter 5

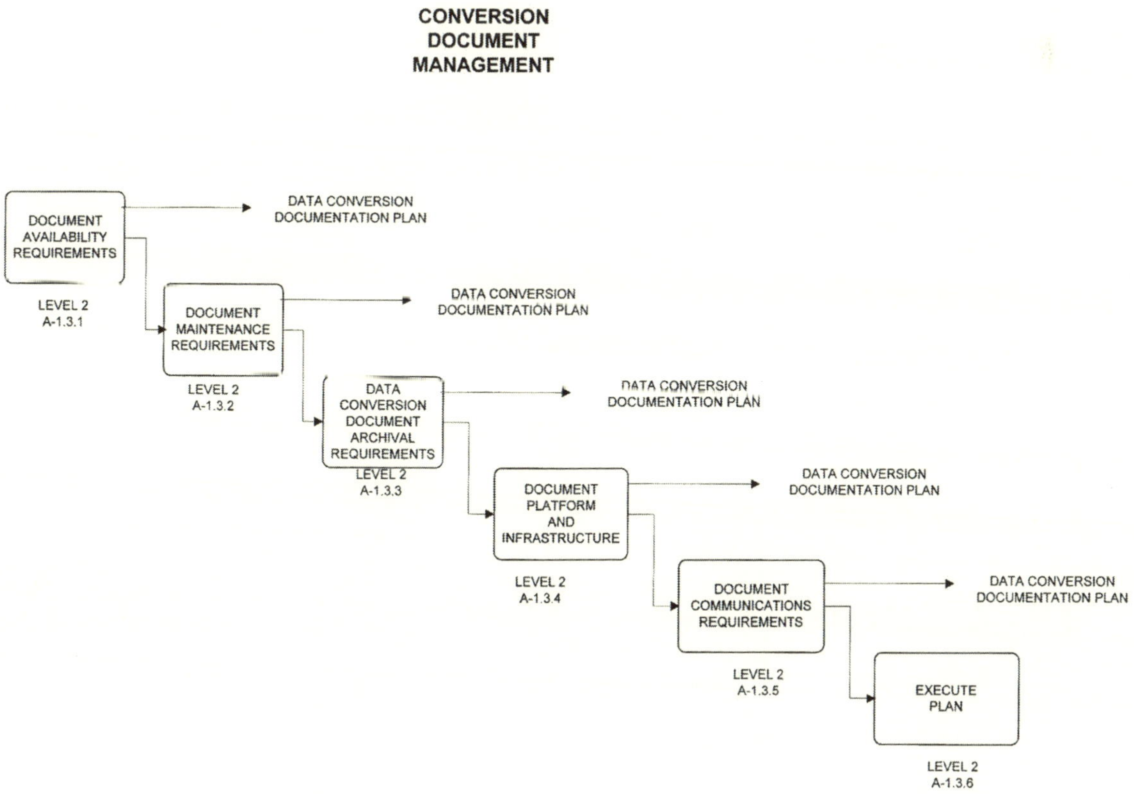

DIAGRAM 4 LEVEL 2 - Document Management

PURPOSE:

The purpose of this section is to provide a document management system to the data conversion team so that all the data conversion documents are available, and can be maintained and referenced. For the best condition, the project team will have already set up a document management system and the data conversion team will use that system as part of its documentation process. If not, here are some things to consider in setting up the system.

A. **COVER PAGE.** Each document should have a cover page that has following information at a minimum.

1. Project Name
2. Level 2 Node Section Name. Typically a project plan will have a management section, a technical section, an education section, a deployment section and a data conversion section.
3. Level 3 Node Name
4. Level 4 Node Name (If applicable)
5. Document Name
6. Date
7. Creator

B. **DIRECTORY.** The directory structure to house the documents should be based around the cover page, hence the project plan and its schedule. This will assist all users in finding and using the documents.

C. **SECURITY.** The security required to support the document management system is to provide a method for keeping the configuration management of the documents in tact. Unauthorized access to documents should already be in place enforced by the organization. This means that the directories should always be accessible in the read only mode.

D. **MANAGEMENT.** The document management system must always have a responsible person assigned to manage the storing and locking of documents. In addition:

a. There should be a person responsible for making sure that each document stored conforms to the corporate or project standard of documentation with the proper headings, footings, and fonts.

b. There should be a method for change distribution. The changes could be as simple as an e-mail to each member on the project team, who then could go review the document for impact on their individual pieces.

LEVEL 2, NODE A-1.3.1 DOCUMENT AVAILABILITY REQUIREMENTS

The requirement for the availability of documents that everyone is working on for the project must be determined. This will establish the ground rules for making the documents available on an as needed basis, and will provide information to all participants on the project. In many cases, it will prevent duplication of effort, saving time. This is where data conversion template requirements, data glossary, and data definitions are established and project plans are available to project personnel.

The need for this node is without question. As directories on servers become available at the individual level, a document, after it is approved is often kept in one of these directories and updated on a continuous basis, with the originator of the document never updating the original in the "sanctioned" directory, simply because there is no procedure for doing such.

Example: (True story)

I can recall having to look for a document in a user's directory for a document that contained a vital data element conversion specification. The responsible person who stored the document was on vacation. There were at least 50 directories, 2,000 sub-directories, and 1,000 sub-sub directories on the server in question. The document was never found during the whole week (quite secure if you ask me). When that person returned from vacation, it took about 4 hours for them to find the document. We lost over 3 days time on an already late project.

LEVEL 2, NODE A-1.3.2 DOCUMENT MAINTENANCE REQUIREMENTS

As documents are changed during the course of the project, a need exist to maintain the documents to the latest revision. Maintenance implies revision control. The requirements for the maintenance and revision control should be determined and included in the process.

LEVEL 2, NODE A-1.3.3 DOCUMENT ARCHIVAL REQUIREMENTS

Archiving documents and having access to documents that are either archived, or reside in another department's documentation system which are project related should be addressed. This may seem unimportant, but in reality, retrieving a previous revision of a document is important in the course of the project.

LEVEL 2, NODE A-1.3.4 DOCUMENT PLATFORM AND INFRASTRUCTURE

A decision must be made as to the platform that the documents will reside on. If in the organization, the document management system is well established and can fit into the requirements of the data conversion project, then it should be used. However, beware that if the current system is very cumbersome and if used, many hours will be added to the project. Once the platform is determined, then the process for retrieving, updating, deleting, and adding documents should be documented. It is very important to make sure that the infrastructure is in place to support this system.

LEVEL 2, NODE A-1.3.5 DOCUMENT COMMUNICATION REQUIREMENTS

Communication requirements of document changes that are made to members of the project team should be documented. This communication is required so that others working on the project can assess the change to see if it has any impact on the work they are doing. This can relate to data field size changes, data definition, transformation rules, etc.

LEVEL 2, NODE A-1.3.6 EXECUTE

A simple documentation system based around the current word processing system may be all that is required. However, the process and procedure to interface with this documentation should be well documented. Other choices are:

- Use the document management system currently in place.
- Use the word processing system currently in place.
- Use the current word processing system that the company has and enhance its functionality.
- Buy an off the shelf system (There are some interesting LAN base systems available).
- Other

PROJECT NOTES

Chapter 6

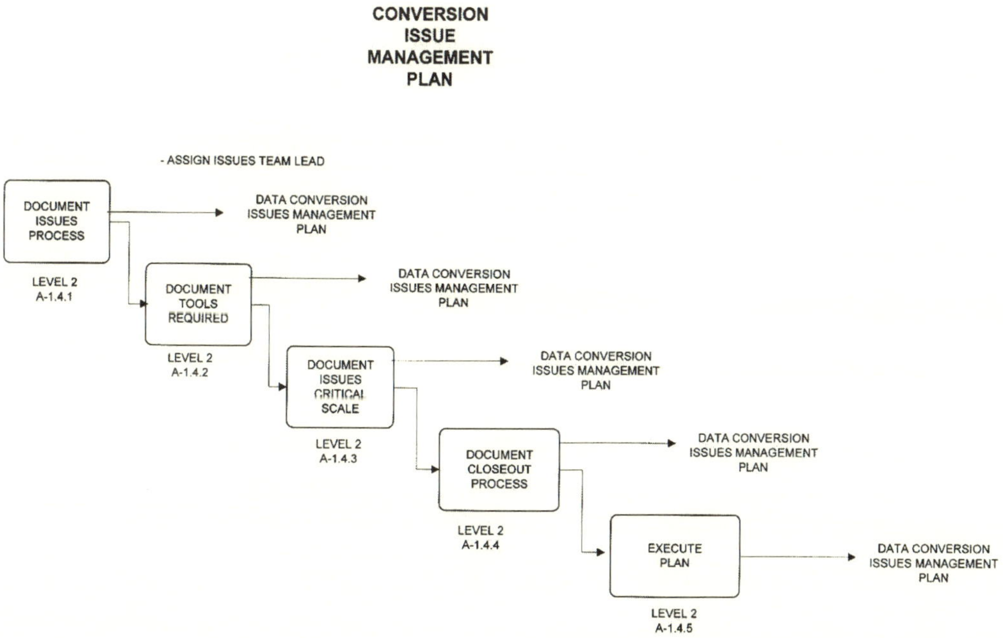

DIAGRAM 5 LEVEL 2 – Issue Management Plan

PURPOSE:

The purpose of the data conversion issues management plan is to provide a sensible plan and an execution process in order to resolve problems and issues that arise during the data conversion process. Too often, the problem resolution process is an afterthought on the project and is started on an ad-hoc basis.

A problem can present itself as a difficulty that arises as a result of any one of the processes in this methodology, or it may present itself in the form of a person. Converting the problems into issues then, makes the problem open for discussion and review.

The first thing that should be done in the process is to assign a lead for the issues log. This person in turn will assist the project manager to set up a steering committee to determine what should be on the issues list. In some instances, one person can make this determination. Some installations make a habit of logging every problem or issue in the issues database and managing them with the category, priority, and status settings.

In most installations, a help desk system for operations is already in place. These systems generally are not good for handling the data conversion issues because of the nature of the problems arising in the data conversion environment, and because of the personnel involved with the help desk system. However, as the date for conversion weekend becomes closer, some of the issues in the data conversion environment may be transferred to the operational help desk system.

LEVEL 2, NODE A-1.4.1 DOCUMENT ISSUE PROCESS

The process for issue management should be a high priority on any data conversion project. The process must specify the beginning of an issue, that is, how to get a problem identified as an issue, to the closing out of the issue or resolving the problem. A process for resolving issues is represented in the diagram of Chapter 17.

LEVEL 2, NODE A-1.4.2 DOCUMENT TOOLS REQUIRED FOR THE ISSUE PROCESS

To manage the issues that arise on an hourly basis in a data conversion environment, tools are required so that issues may be logged, sorted and printed by priority, category, owner, and many other different ways. In addition, some issues require a sequential process to resolve. That is, user A will have to supply some information for a solution, then user B, then user C. The tool you select for issues management process must be able to handle this situation.

LEVEL 2, NODE A-1.4.3 DOCUMENT ISSUES CRITICAL SCALE

Another important criterion for managing issues is a determination of the criticality of the problem. Once the issues have been categorized with a critical number, the most important problems can then be assessed for their relevance in getting the data conversion completed. A scale should be consistent with the thinking of the rest of the data conversion scales. For instance the scale may be as follows:

Critical Issue = 1

An issue designated as critical signifies that the associated task cannot move forward until the issue is resolved. If there are dependencies associated with the task, then this could impact the conversion schedule in a major way. If necessary, when an issue is designated as critical, realign resources, both man/woman and machine to resolve the problem. Critical means that this issue has major importance to the completion of the project on schedule and must be addressed immediately.

Major Issue = 2

A major issue is one that will have an impact on production if not resolved prior to production cut-over weekend. This could relate to a task that, due to its complexity, has been discovered will take longer than anticipated to complete as per the schedule. The relating issue should be commented to reference the delay. If this task is not completed as per the scheduled due date of the issue, then it must be moved to a critical issue.

If dependencies are linked to the issue, then there must be a work a-round specified that will assure the data conversion effort can precede until the issue is resolved.

Minor Issue = 3

A Minor issue is considered to be desirable when it will have no impact on the data conversion efforts, or impact the system when production starts but should be corrected. This type of issue may be handed over to the help desk for resolution and in turn, may get a higher priority.

LEVEL 2, NODE A-1.4.4 DOCUMENT CLOSEOUT PROCESS

Closing out an issue in the issues management system can sometimes be a little tricky, especially if those responsible for the issue are not easily located. It is recommended that for critical issues, a daily meeting be held to discuss the solution until the issue is resolved. For major issues, meetings of once a week should be conducted for the same reason. Each issue resolution should have the appropriate comments that specify the solution.

It is possible that some issues generate more issues. If this happens, make sure the tool can keep track of the relationships. Also, make sure that issues generated are not delay tactics (unintended of course). "Parking Lot" items that come out of issues meetings can become a snag to the data conversion process if not attacked with vigor.

The close out of an issue should leave no doubt in the managements mind that the issue is closed. If doubt exists, leave the issue open until the air is cleared.

LEVEL 2, NODE A-1.4.5 EXECUTE

Make sure a responsible person is in charge of the tool and the process to manage the issues that arise on a data conversion project. Conduct regular meetings on a scale to resolve all issues in a timely manner, and make sure issues are assigned to responsible people.

Chapter 7

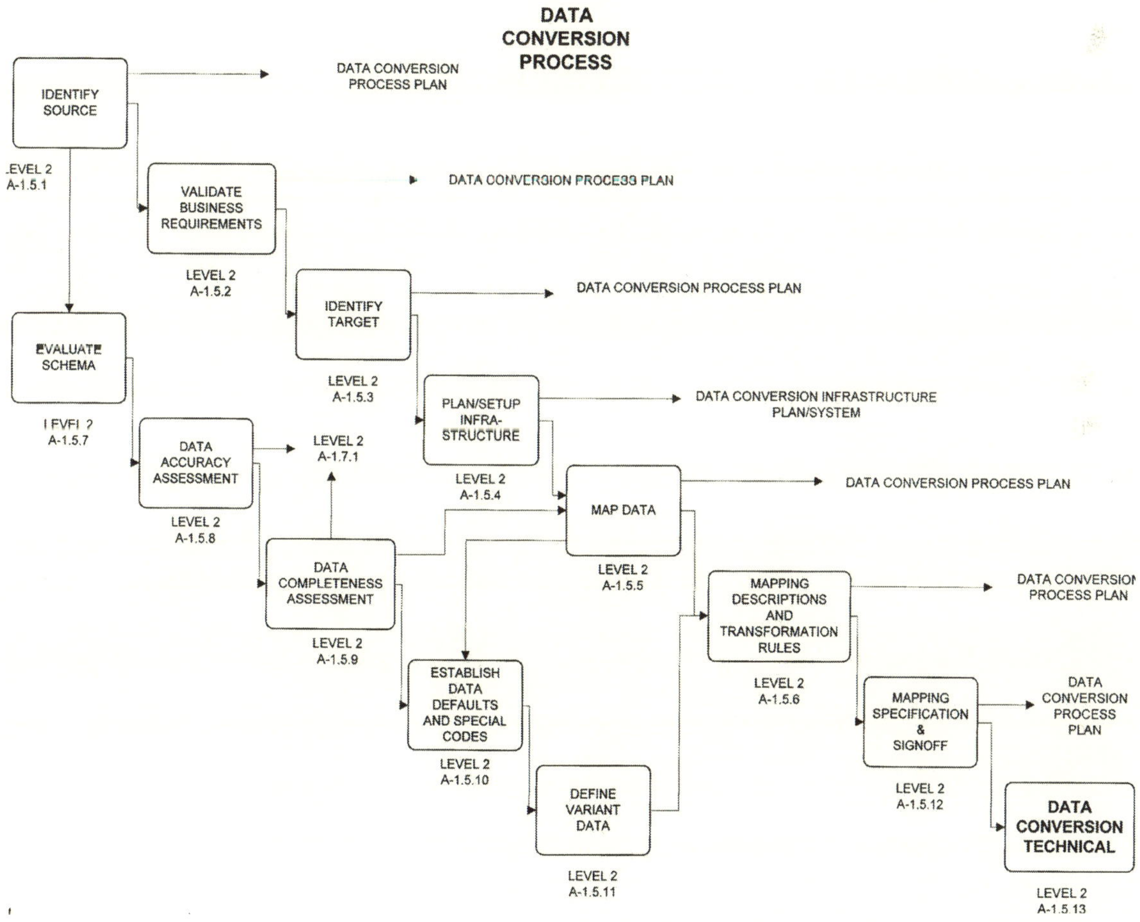

DIAGRAM 6 LEVEL 2 - Data Conversion Process

PURPOSE:

The purpose of this section is to describe the data conversion process that will be used to transform and convert the data. This process can be the most difficult to manage and communicate in this methodology. To avoid confusions, the previous steps in this methodology serve as a predecessor. At a minimum:

- Have the scope defined
- Have the data ownership identified
- Have the topography of the data completed
- Determine what data is to be archived for future reference versus converted for current operations

LEVEL 2, NODE A-1.5.1 IDENTIFY SOURCE

The source data identified in the TOPOGRAPHY node of A-1.1.2 is now analyzed in detail for the data that is needed for the conversion process. This implies that some knowledge of the "new" system is required and that complete knowledge of the legacy system is mandatory. In other words, you must identify the source of data for the application that you are converting. All the sources must be identified that are within the **scope** of the conversion.

LEVEL 2, NODE A-1.5.2 VALIDATE BUSINESS REQUIREMENTS

Once the source is identified, a validation must be made of the source to make sure that all the data identified for conversion will meet the business requirements. This includes all "history" data and archived data. Any data left behind that is needed for the business must be evaluated and bought to the attention of those in charge.

It is most important at this stage to include the "CONTROLS" part of the project evaluation of the business requirements. If there are state laws, federal laws, or some agency laws which require data but are not part of the business process, or if the application requires data that is not part of the business process, then this information must be included as part of the conversion process.

This is also a good time to assure that business requirements that have evolved are included. This often requires reviewing special management reports that are part of the legacy system. In addition, these special requirements should be mapped against the new system to make sure they can be met.

LEVEL 2, NODE A-1.5.3 IDENTIFY TARGET

For the conversion to be performed, the target tables, files, or segments must be identified that are required for the application. If the application is new to the company, the target may be unfamiliar and therefore, some of the fields required by the target or application but not required by the business may be missed. It is always a good idea to hire someone familiar with the application vendor company to help with the mapping. It certainly is worth the cost. There are just too many cases where data has been converted, the application started, and transactions ABEND because of an incomplete, or a missed data field.

It must be understood that in most cases, the target tables will have the data organized completely different from any legacy system currently in operation. In some ways, it may not make any sense to the technical resources performing the mapping and providing the transformation rules, especially if that technical support has been associated with the legacy system for the past 10 to 20 years (as is the case most of the time).

LEVEL 2, NODE A-1.5.4 PLAN AND SET UP INFRASTRUCTURE

The infrastructures to extract, transform, and move the data to the target must be planned and set up. It must be established from a project viewpoint so that all the technical resources working with the data can adhere to the process. Otherwise, everyone will have their own way of moving and transforming data, creating inefficiencies in the project. The infrastructure should also consider "conversion weekend". If lots of data has to be moved or transferred, and conversion to production time is short, the infrastructure must be able to perform in the conversion window.

Any new tools required to set up the infrastructure should be identified and purchased. Usually these tools are not very expensive and should be purchased as soon as they are identified. Special written programs may be required to set up the infrastructure.

Whatever you do, make sure that personnel are available to support the infrastructure. This could be in the form of network support, software support, hardware support, or DBA support. It could be as simple as having a member of the conversion team take on additional tasks.

LEVEL 2, NODE A-1.5.5 MAP DATA

Mapping of the data is probably a term most associated with data conversion. The source data is mapped to the target fields. Mapping is simply a process of selecting which data from the tables or files of the legacy system must be converted to the tables or files of the new system to meet all the business requirements. If there is data in the legacy system required for the business processes that have no apparent home in the new system, then either a home is found from another data field, or new data fields have to be added to a table. In either case, the data dictionary should be updated which will require the beginning of configuration management for the data, especially if the new system is a vendor off the shelf package and new fields are added to the tables that are not vendor supported.

Concurrent activities with this node are nodes A-1.5.7 to A-1.5.9.

NOTE:

If you are a member of a data conversion team, and the first task is to map the data, you must consider the prerequisites and should have reviewed the data conversion project schedule to understand where you fit in the big picture, reviewed the requirements document, been introduced to the user data interface for the legacy system, and provided with a mapping template. Of course one can fake it and pretend to know what one is mapping until one gets all the answers needed which is sometimes the case. This happens because the prerequisites do not exist and legacy data knowledge is very limited.

LEVEL 2, NODE A-1.5.6 MAPPING DESC AND TRANSFORMATION RULES

After the data is mapped, each mapped field should be described and the transformation rule (if required) provided to meet the requirements of the target data field. It is at this point that rules are created for data that is required for the new system, but are not in the legacy system. Concurrent activities with this node are nodes A-1.5.10 to A-1.5.11.

LEVEL 2, NODE A-1.5.7 EVALUATE SCHEMA

The schema for the source data must be available. It in fact must be available before transformation rules are completed. The schema represents to data what the drivers' license represents to the individual. It tells the sex (data format), weight (size), address (source table), etc. of the data. The schema is needed to determine if a transformation rule is needed to satisfy the target and will give some guidance in the assessment of the data.

For example, if the source is a numeric field of six (6) positions with three (3) decimal places and the target is numeric of ten (10) with six (6) decimal places, then a transformation rule has to be written to make the conversion. If the source schema is not evaluated for these differences, then the data will not go across accurately.

LEVEL 2, NODE A-1.5.8 DATA ACCURACY ASSESSMENT

The data in the source should be assessed to determine the level of accuracy. If the system has had heavy user interface, or if the system is very old, then there is a high likelihood that the data is not as accurate as desired, and will fail to meet the business needs. As part of the data conversion process, the inaccurate data must be cleaned or discarded.

To do this, a process must be in place to determine the accuracy. Some common data areas to review for accuracy are:

- Abbreviations
- Dates
- Special codes

LEVEL 2, NODE A-1.5.9 DATA COMPLETENESS ASSESSMENT

The data in the source should be assessed to determine the level of completeness. Are there blank fields where there should be data? Are there data fields with only partial data? Will converting the data as is lead to misunderstanding in the new system? These questions must be answered prior to turning the programmers loose on programming.

LEVEL 2, NODE A-1.5.10 DEFAULTS AND SPECIAL CODES

Any defaults or special codes required for the new system must be established. Many off the shelf packages have provisions for these defaults or require special codes or setup parameters. These must be established so that they can be considered before the mapping specifications are finalized.

LEVEL 2, NODE A-1.5.11 VARIANT DATA

Variant data is data from the source or legacy system that will be required in the new system, but has a different meaning than the target field description. This may also be described as a placeholder for the source data or be used as a means of referencing data.

LEVEL 2, NODE A-1.5.12 MAPPING SPECIFICATIONS AND SIGNOFF

The result of the data conversion process is a mapping specification that can be turned over to Data conversion technical resources to write the programming specification. In most cases, the mapping specification will be sufficient for the programmer to write the code to extract and transform the data for conversion. At any rate, sign off of the specification must occur for approval.

PROJECT NOTES

Chapter 8

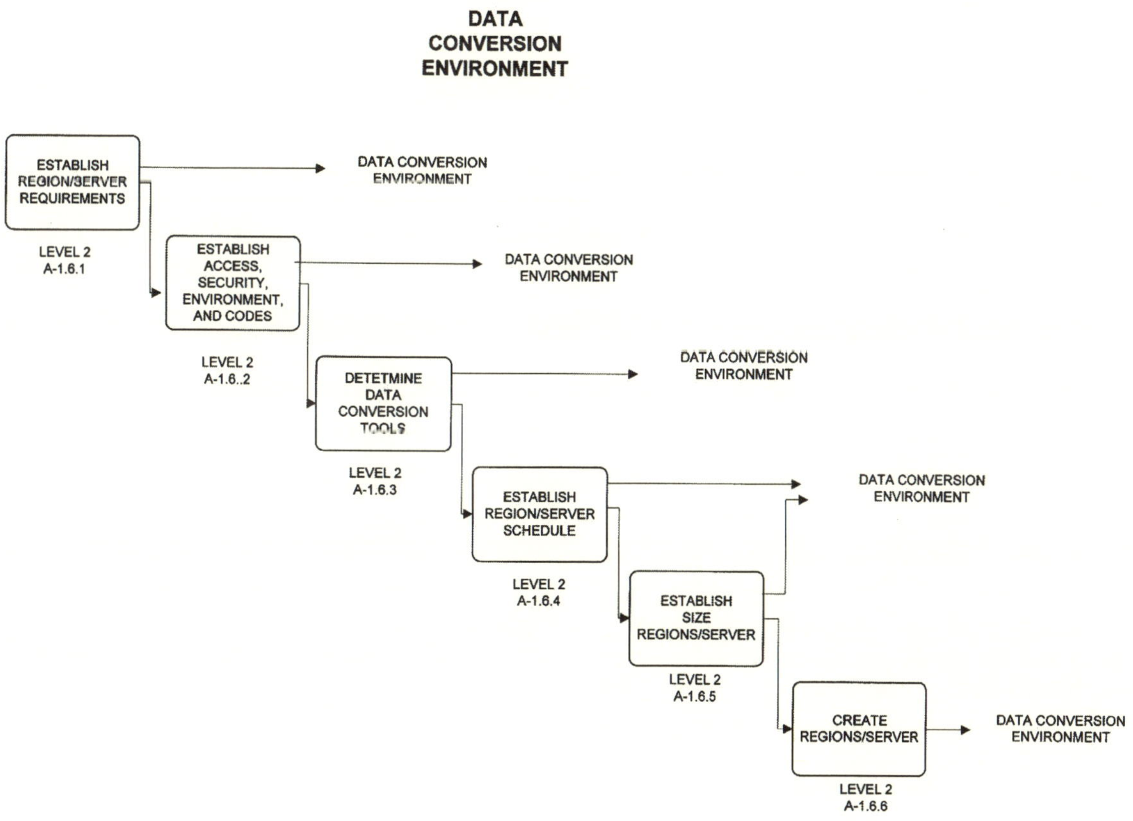

DIAGRAM 7 LEVEL 2 -Data Conversion Environment

PURPOSE:

The purpose of this section is to describe the tasks that should be considered in setting up the data conversion environment to successfully convert data and save time and dollars. This environment defines the parameters from which work can be accomplished at the physical location. Appoint an Environment Administrator to manage the activities within this level.

LEVEL 2, NODE A-1.6.1 ESTABLISH REGION/SERVER REQUIREMENTS

To meet the needs of the data conversion team, and to meet the data needs of the project, a place is needed in the hardware/software environment to avoid confusion when determining where to perform various tasks. It is not uncommon during the data conversion process, to have multiple environments to support the process. These environments are generally called regions. Normally a region is needed for each of the following areas with data, whether new sample data or new converted data to assist in the testing and execution:

- unit testing of programs written for conversion (data extracts and transformations)
- Unit testing of data load programs
- System testing of data load programs
- Acceptance of loaded data in new application
- Performance testing

In addition, other regions must be considered for the project such as:
- User training on the new application
- New application code fixes and testing

These requirements must rest on top of the current operational system which places an additional burden on the system and those that must support both environments.

These environments must be established and availability planned in a timely manner.

LEVEL 2, NODE A-1.6.2 ACCESS, ENVIRONMENT, AND SPECIAL CODES

Access to each of the planned environments must be determined, based on the need to access the data within the environment. The application users, in coordination with the IT environment administrator will manage who gets to perform what functions in which environment. The administrator will also have to work together with the Systems Programmer, LAN Administrator, and Data Base Administrator to ensure the environment is established and parameters set to meet the needs of the project.

LEVEL 2, NODE A-1.6.3 DETERMINE DATA CONVERSION TOOLS

As part of the environment setup, all tools required to complete the data conversion, or to assist in the conversion of data must be identified, purchased, and installed to function in the environment. The tools needed can be derived form the topography of the source data, and the target data and should have been identified as a result of performing LEVEL 2, NODE A-1.1.5, - DATA CONVERSION DESCRIPTIONS.

LEVEL 2, NODE A-1.6.4 ESTABLISH REGION/SERVER SCHEDULE

To assure that the environment is available when required, the events in the schedule that require one of the environments must be back scheduled to assure the environment is ready when needed. As dates in the schedule change, all prerequisites should be highlighted.

LEVEL 2, NODE A-1.6.5 ESTABLISH – SIZE REGION/SERVER

Based upon the data needs of the environment, and the number of users, each environment must be sized for data space, user retrievals, and module execution.

Charles R. Scott

More often than not, on a data conversion project, there is never enough data space and the computer processor is never fast enough.

LEVEL 2, NODE A-1.6.6 CREATE REGION/SERVER

As needed, since the environment is planned, create or implement the environment on the computer. When conflicts arise, reset the schedule to reflect changes as they surely will happen.

Chapter 9

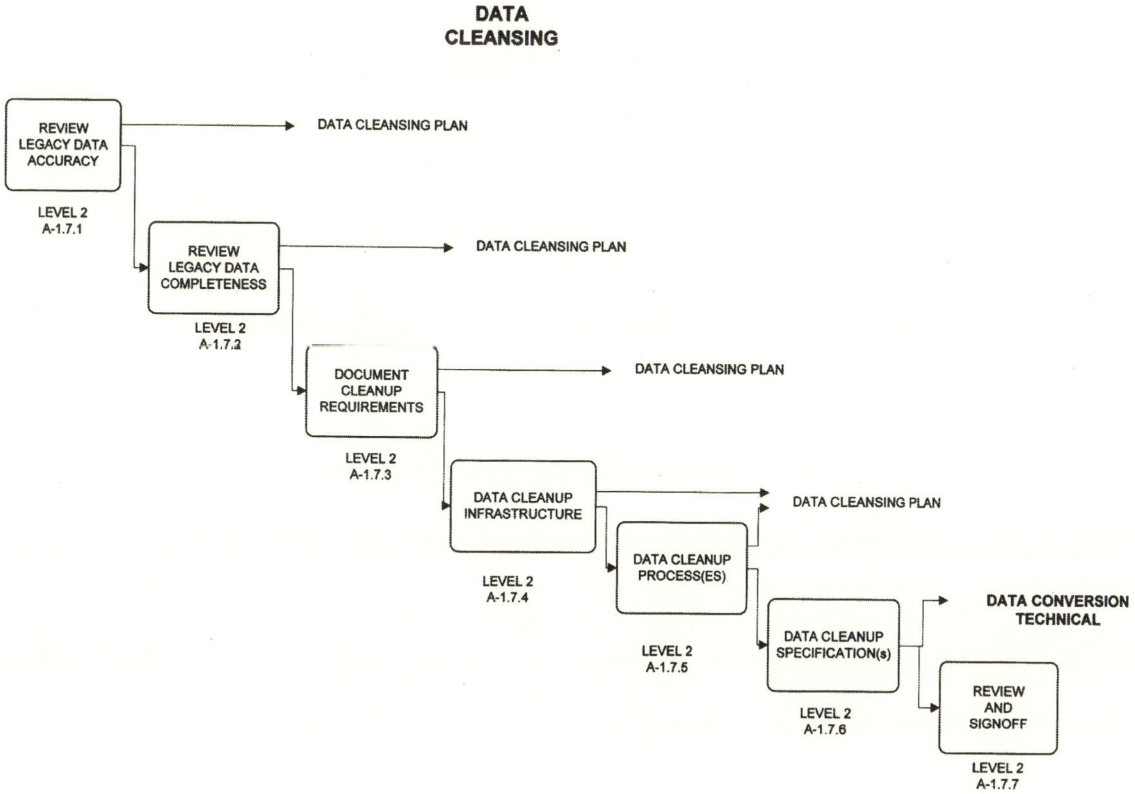

DIAGRAM 8 LEVEL 2 - Data Cleansing

PURPOSE:

The purpose of this section is to assure that data converted as part of the implementation project is cleansed. This section may be entered as part of the Data Conversion Process, provided the cleaning activities are not extensive. However, if the data cleanup activity is extensive, establish a separate or side project to perform data cleansing. Data cleanup may take several forms. It may consist of manually collecting data from a legacy source and correcting it, manually collecting legacy data and comparing it to an automated source, cleaning up data from a legacy automated source, or any combination of the above.

When considering whether to cleanse the data or not, always remember the old adage of "garbage in equals garbage out". If data is worth storing, it is worth being stored correctly.

Cleansing data also provides authenticity of the new system to the end users of the data. When the end user is using an on-line transactions and can "trust" the data, then the system will be used to its fullest.

LEVEL 2, NODE A-1.7.1 REVIEW LEGACY DATA ACCURACY

This node is entered from nodes A-1.4.8 and A-1.4.9. After the data has been collected and analysis performed to determine the accuracy levels of the data, a review must be conducted to establish time/value for cleaning the data. This review validates accuracy requirements as it relates to integration of the various applications modules that are to be implemented. Again, some common data areas to review for accuracy are:

- Abbreviations
- Dates
- Special codes
- Status codes

- Interpretative specific data
- Numbers (Order numbers, Inventory numbers, etc.)

LEVEL 2, NODE A-1.7.2 REVIEW LEGACY DATA COMPLETENESS

After the data has been collected and analysis performed to determine the completeness levels of the data, a review must be conducted to establish time/ value for completing and filling in empty holes in the data. This review validates completeness requirements as it relates to integration of the various applications that are to be implemented.

Completeness of data relates to empty data fields or spaces in a data field that relates to a hierarchy of data. For instance, in the name and address file, are all the addresses filled in? Are all the states filled in and correct? Are all the zip codes filled in and correct? This of course, is a simple example and data completeness and may have no value to the operational considerations of the application. This must be assessed. However, even if it has very little value, it is worth the time and effort to clean this data.

LEVEL 2, NODE A-1.7.3 DOCUMENT CLEANUP REQUIREMENTS

Data field cleanup requirements are documented. How and when the cleanup will occur will be part of this process. In some cases, the cleanup map occurs as part of the extract and transformation of the data in the conversion process. Other cases may demand that a separate process for cleaning the data occur outside of the data conversion stream and before cut over weekend, or it could be a separate process that is carried out after cut over weekend.

LEVEL 2, NODE A-1.7.4 DATA CLEANUP INFRASTRUCTURE

The infrastructure to clean up data should have been included in node A-1.3.4 which included considerations for the topography of the data. Finalize and set up as to how the data will be moved from, or entered into the cleanup system, as well as the tools that will be used. It may be that some in-house programs will be written to clean up the data. These need to be identified.

LEVEL 2, NODE A-1.7.5 DATA CLEANUP PROCESS (ES)

Describe and flow the process for cleaning up the data to obtain the desired results. This cleanup process must have quality steps embedded to assure that the final data will meet the stated requirements. Again, this process may be executed before the data conversion week end, during the data conversion weekend, after data conversion weekend, or a combination of these activities.

LEVEL 2, NODE A-1.7.6 DATA CLEANUP SPECIFICATION

The output should be a data cleanup specification. This specification need not be another version of the dictionary. At a minimum it should contain:

- A specification number (This may relate to an issue statement or a data defect number)
- An objective of the specification
- A description of what has to be done
- A SQL statement of the code required.

However, it should be definitive, even if only one paragraph long, to obtain the desired results.

LEVEL 2, NODE A-1.7.7 CLEANUP SPEC REVIEW AND SIGNOFF

The cleanup specification must be approved and signed off by the responsible data owner and subject matter expert. After signoffs, the specification can be sent to the technical representative to perform the data cleanup activities. This representative may be a member of the user community, or someone that is familiar with the tools and infrastructure to clean up the data.

As with any clean up activity, there should be a server region associated with the data clean up activity so that adequate testing and Q/A can be performed prior to implementing the clean up activity in production.

PROJECT NOTES

Chapter 10

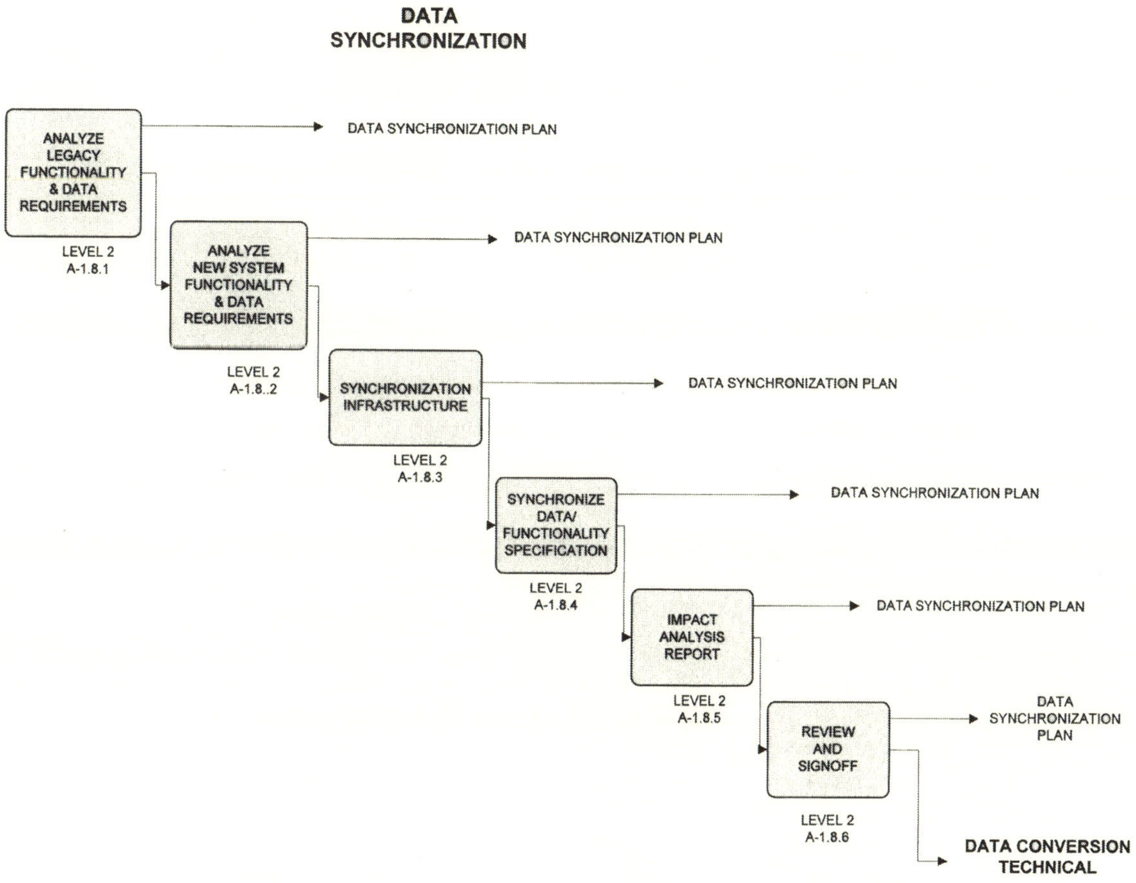

DIAGRAM 9 LEVEL 2 - Data Synchronization

Charles R. Scott

PURPOSE:

The purpose of this section is to provide and plan for the coordination of data that may be left behind in the legacy system with the data that is converted to the new system. As active transactions may reside in both systems, it is required to synchronize the data between the two systems. The solution may serve as a temporary one until all the systems have been converted, or it may be a permanent solution where two systems are designed to work in tandem. In any case, where data overlaps in the two systems, synchronization must occur.

LEVEL 2, NODE A-1.8.1 LEGACY FUNCTIONALITY & DATA REQ'MENTS

The legacy functionality or data that is not scheduled for archiving, and needs to interact with the new system or new functionality needs to be analyzed. In some cases, there may be an overlap in data that will need tight synchronization when determining the timing of information retrieval from the two systems.

LEVEL 2, NODE A-1.8.2 NEW FUNCTIONALITY & DATA REQUIREMENTS

The new system functionality or data that is scheduled for implementation and needs to interact with the legacy system or old functionality needs to be analyzed. In some cases, there may be an overlap in data that will need tight synchronization when determining the timing of information retrieval from the two systems.

LEVEL 2, NODE A-1.8.3 SYNCHRONIZATION INFRASTRUCTURE

The infrastructure to determine how the data will move between the two systems, the timing of the move, and the updating of the data must specified and tested. All these factors are important and must be optimized for efficiency of operations and optimization of system performance.

A common trend in today's hardware/software environment is to use trigger mechanisms in each environment that captures batch transactions for processing. If this type of mechanism is used in the environment, make sure the infrastructure is in place to support this type of processing.

LEVEL 2, NODE A-1.8.4 SYNCHRONIZE DATA/FUNCTIONALITY SPEC

A specification will be written to spell out exactly how the synchronization or interface will work. It will include the data elements involved, how the data will be processed, target fields, how the data will move over the infrastructure, and the timing of the process.

LEVEL 2, NODE A-1.8.5 IMPACT ANALYSIS REPORT

An impact analysis report should be prepared at this point to determine the risks associated with legacy data retention versus archiving.

LEVEL 2, NODE A-1.8.6 REVIEW AND SIGNOFF

As with every level 2 activity, there must be a sign-off and approval process that permits work to proceed in getting the programs written and tested to implement into production with the rest of the system. The approved specification goes to the technical group for execution.

PROJECT NOTES

Chapter 11

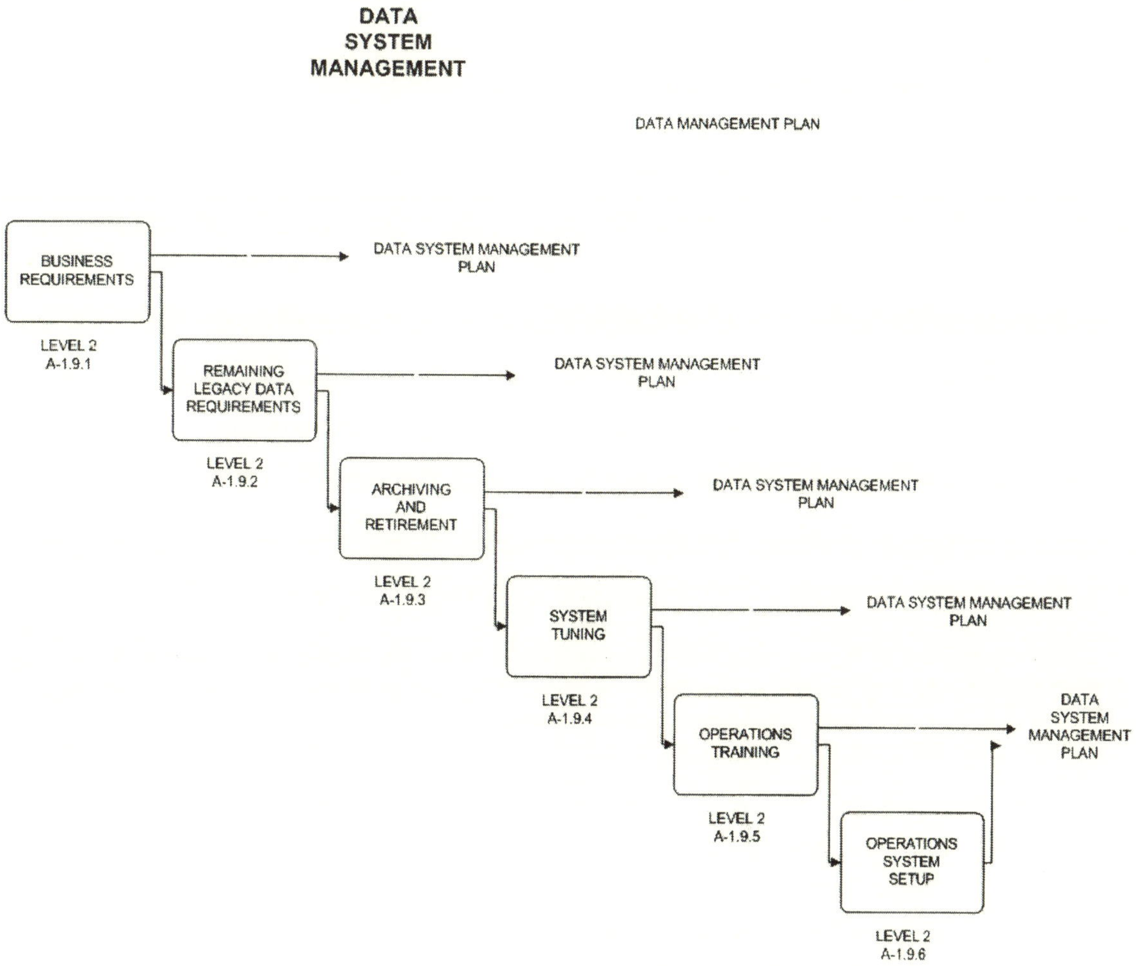

DIAGRAM 10 LEVEL 2 - Data System Management

PURPOSE:

The purpose of this section is to provide and plan for the implementation of the new software, database, or application into the production environment. This activity should be considered and planned as part of the overall project schedule and its activities scheduled with the data conversion activities.

LEVEL 2, NODE A-1.9.1 BUSINESS REQUIREMENTS FOR PRODUCTION

The business requirements for the company should be documented as part of the new system implementation. These requirements include technical support, application help desk support, transaction response times, batch run time schedules, etc. Document these requirements as part of the data conversion project to insure that they have been addressed and can be met.

LEVEL 2, NODE A-1.9.2 REMAINING LEGACY DATA REQUIREMENTS

The legacy system that will remain in production cannot be ignored. Users by this time usually have some expectation of the capabilities of the legacy system and will not expect less in the future as the new system goes on-line. Be sure that processing conflicts do not arise as a result of the new system.

LEVEL 2, NODE A-1.9.3 ARCHIVING AND DISCARDING

A decommissioning plan for the legacy data must be established and executed as the new system goes on line. Generally, until the new system is in place, and for a time period of 30 to 180 days thereafter, the legacy system remains on-line in a display mode only. This is often necessary to support production until the new system becomes a way of life for the users.

LEVEL 2, NODE A-1.9.4 SYSTEM TUNING

There are many elements that go into system tuning. Perhaps the system will not need tuning with the new application because you have a fast computer with a lot of memory, applications that are optimized for speed, and a database that is perfectly related. However, for those that do not, it should be determined if the system needs tuning. If so, then how. DBA's are generally a great first line source for tuning.

LEVEL 2, NODE A-1.9.5 OPERATIONS TRAINING

Oftentimes, the users are trained on the new system, but operations personnel are neglected until the week that the new systems are to go in production. In reality, new operational procedures and training must be in place before production is successfully implemented.

LEVEL 2, NODE A-1.9.6 OPERATIONS SYSTEM SETUP

Procedures, job streams, timing of execution, backup procedures, and database reorganization are just some of the setup steps that need to be accomplished prior to the production week. Attend to these activities with gusto.

PROJECT NOTES

Chapter 12

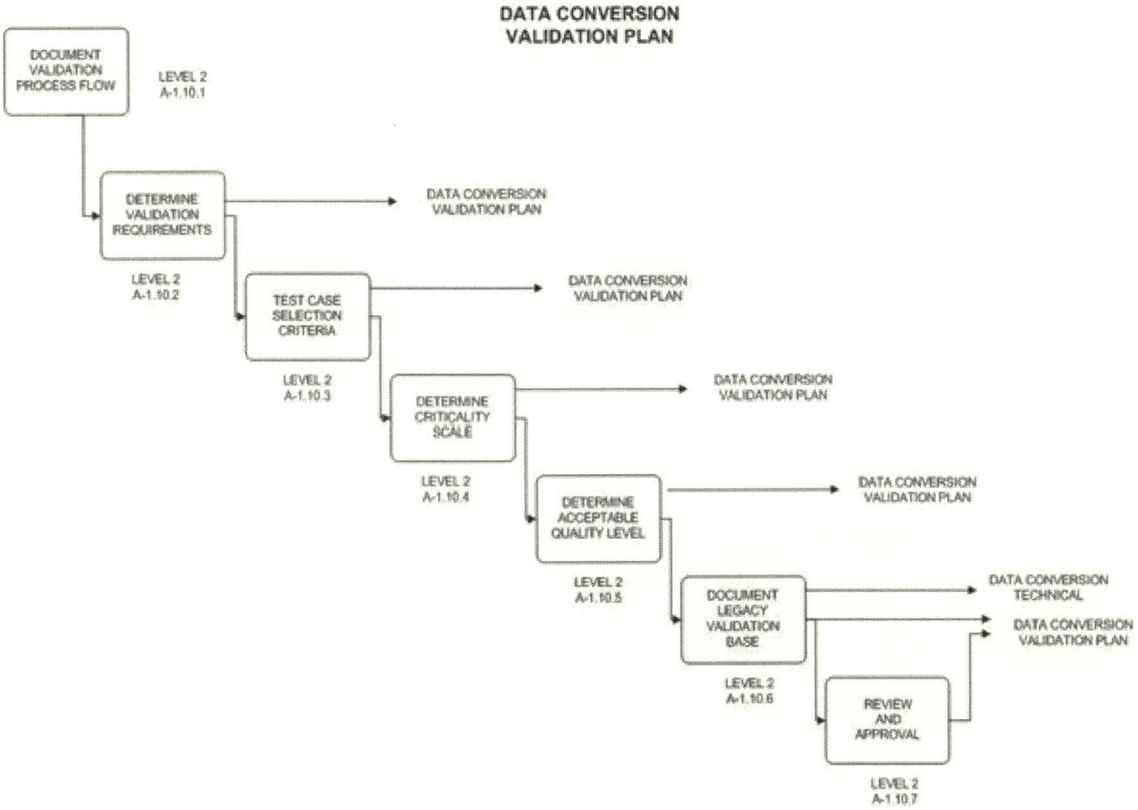

DIAGRAM 11 LEVEL 2 - Data Validation Plan

PURPOSE:

The purpose of this section is to plan for the validation of the converted data. This validation will take place before the user performs the business process data validation and is part of the whole data conversion effort. It must be performed each time a test conversion takes place to assure the data is being converted properly.

LEVEL 2, NODE A-1.10.1 DOCUMENT VALIDATION PROCESS FLOW

The process flow from the beginning of the data conversion process through the validation execution must be documented so that those involved with the validation will have a clear picture of what has occurred, and how the desired results will be obtained. Each step of the validation process should have an approval, or must be performed to a documented requirement.

The process flow for validation may look something like the following diagram.

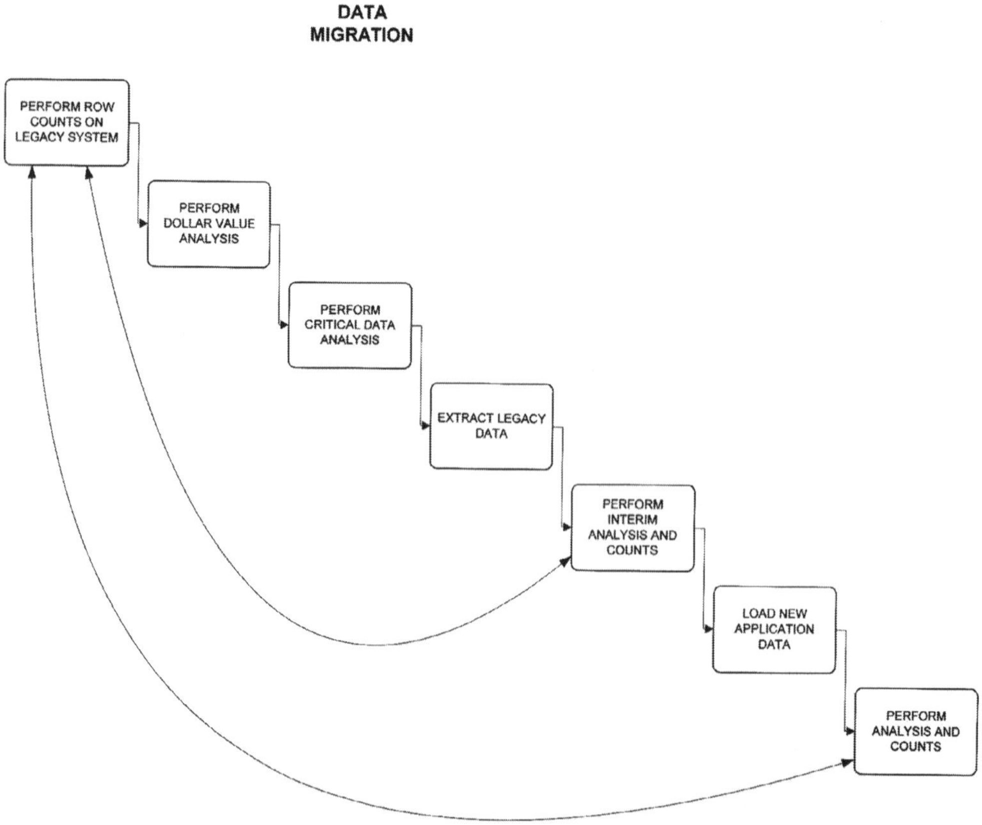

Once documented as the process, the steps may be performed many times. One thing to keep in mind is that this process has to be performed during conversion weekend. Therefore it is in the best interest of the project to streamline this process as much as possible. If the process is too cumbersome, it will extend the weekend by a few hours and make some expectant users, which should be avoided at all cost.

LEVEL 2, NODE A-1.10.2 DETERMINE VALIDATION REQUIREMENTS

In planning for the validation of the data, all requirements must be established. These requirements include business requirements, cleanup and transformation rules, new application requirements, defaults, special codes, and processing requirements.

LEVEL 2, NODE A-1.10.3 TEST CASE SELECTION CRITERIA

When selecting the data to be validated, all general and any special criteria must be included in the validation. The expert users, the subject matter expert, and/or the data owner can provide the selection criteria. Records should be selected so that all the ifs, ands, and ors can be tested.

Test case selection can be based on legacy record types like work order types, inventory types, document types, purchase types, or accounting types. If there are multiple data types within an application system, then each type must be validated.

LEVEL 2, NODE A-1.10.4 DETERMINE CRITICALITY SCALE

A level of criticality should be established for data elements during the validation phase. Sample levels of criticality are shown below:

Critical Defect

A critical defect of converted data is a defect that renders the converted data useless. A critical defect relates to the critically of the application to the business process. There are many examples of this during a conversion process and each must be documented prior to conversion. It is possible that critical defects are not identified prior to production in which case, production on the new system will be halted until the corrections are made.

Critical defective data must not be converted and must be corrected prior to production.

Major Defect

A major defect of converted data is a defect that results in the reduced usability of the converted data.

Major defective data **must** have either a work around or a corrective plan in place and enforced at production time that satisfies the business requirements. Since many of the vendor ‘off the shelf” software packages do not always provide an exact duplicate for legacy system transactions, these failings are often circumvented by some work around method. These work around methods are usually documented in the business process flow.

If a major defect relates directly to the data, then the problem and solution of the defect should be in place prior to production and reinforced when the system is put into production.

Minor Defect

A Minor defect of Converted data is a defect that will have little bearing on the effective use of the data.

Minor defective data **may** have either a work around or a corrective plan in place and enforced at production time that satisfies the business requirements. Minor defects will not affect the usability or reduce the effective use of the system. Minor defect generally are never addressed in most application systems, especially in legacy systems that have reached their decay stage. However, it is highly recommended that all minor defects be fixed in the new application system. Remember "garbage in = garbage out".

LEVEL 2, NODE A-1.10.5 DETERMINE ACCEPTABLE QUALITY LEVEL (AQL)

The difficult process in determining the acceptable quality level for data is the mix of criticality of data elements within a data conversion process. According to MIL-STD-105E:

"A sampling plan and an AQL are chosen in accordance with the risk assumed. Use of a value of AQL for a certain defect or group of defects indicates that the sampling plan will accept the great majority of the lots or batches provided the process average level of percent defective (or defects per hundred units) in these lots or batches be no greater than the designated value of AQL. Thus, the AQL is a designated value of percent defective (or defects per hundred units) for which lots will be accepted most of the time by the sampling procedure being used. The sampling plans provided herein are so arranged that the probability of acceptance at the designated AQL value depends upon the sample size, being generally the higher for large samples than for small ones, for a given AQL. The AQL alone does not identify the chances of accepting or rejecting individual lots or batches but more directly relates to what might be expected from a series of lots or batches, provided the steps indicated in this publication are taken. It is necessary to refer to the operating characteristic curve of the plan to determine the relative risks."[4]

For this reason, in a data conversion process where you may have a mix of the criticality of data within the data record being loaded, it is important to have established the AQL for each possible defect within the record.

LEVEL 2, NODE A-1.10.6 DOCUMENT LEGACY VALIDATION BASE

Document all the records that will be selected based on the selection criteria, and the random data sample size required by the Acceptance Quality Level (AQL) documentation. Plan on getting a data report, or screen printouts immediately preceding the validation run. If programs need to be written to produce the reports, write the program specifications and complete the programs so that they can be ready by the time the validation execution occurs.

LEVEL 2, NODE A-1.10.7 REVIEW AND APPROVAL

The subject matter expert, the data owner, and a member of the quality group must approve the validation plan. Once approved, all subsequent activities must then be planned for execution.

Chapter 13

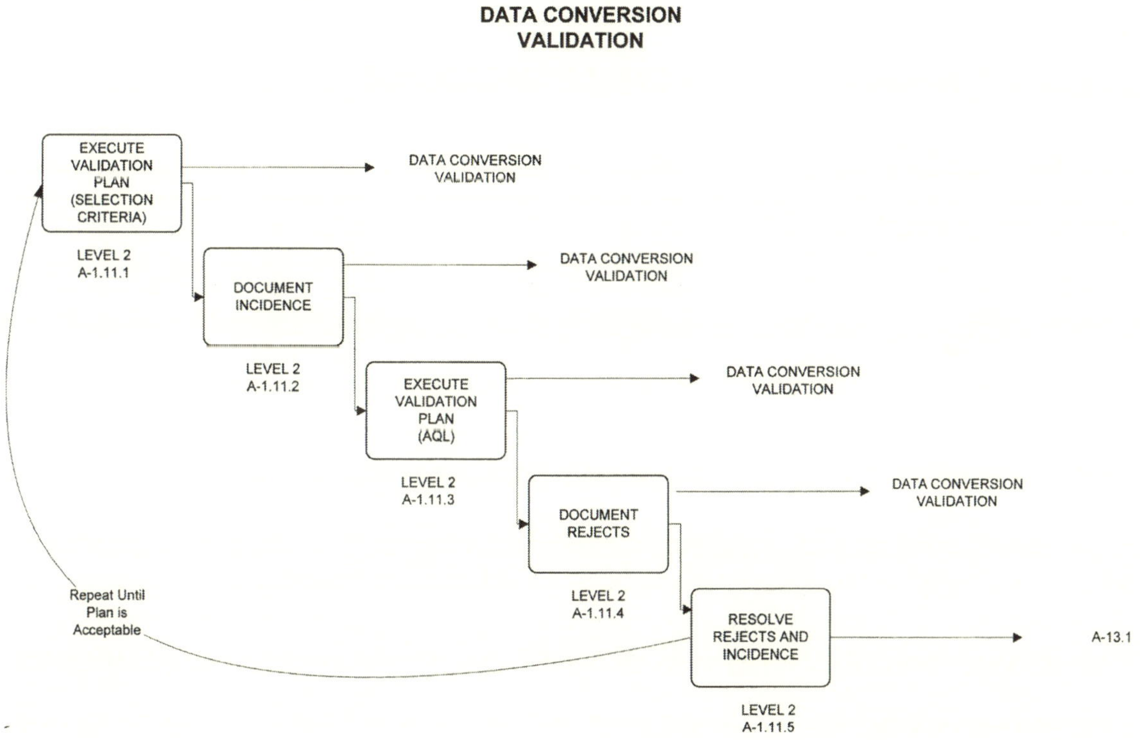

DIAGRAM 12 LEVEL 2 - Data Conversion Validation

PURPOSE:

The purpose of this section is to execute the validation of the converted data according to the validation plan. It is best to keep the approach to validation as simple as possible, while at the same time, obtaining the desired results of the validation requirements.

LEVEL 2, NODE A-1.11.1 VERIFY VALIDATION PROCESS FLOW

Prior to the validation process, verify the validation process flow. Included in the process should be row counts, data element validations, special codes validation, numbers validation (basic numeric numbers or dollars and cents), and data logic validations.

Based on the records and data fields selected for validation, validate the data against the legacy system data and other associated data.

LEVEL 2, NODE A-1.11.2 DOCUMENT INCIDENCE

Any data that does not pass the validation requirement must be documented as an incidence (of error).

The before condition (legacy system), the expected results, and the after condition (new system) must be recorded. All three conditions must be documented and presented to the validation team (which has authorized decision making users involved).

LEVEL 2, NODE A-1.11.3 EXECUTE VALIDATION PLAN

The random selection of records to be tested based on the sample size to meet the Acceptable Quality Level must be executed. Included in this test may be random records based on several data field characteristics whose sum of records meet the criteria of the AQL (Reference MIL STD 105E). In addition, classification of data fields with the criticality scale will help make decisions easier at the "go"/ "no go" point.

LEVEL 2, NODE A-1.11.4 DOCUMENT REJECTS

Rejects identified in the random test for the AQL validation must be documented. The before condition (legacy system), the after condition (new system), and the expected results must be recorded. All three conditions must be documented along with an explanation of any rejects.

LEVEL 2, NODE A-1.11.5 RESOLVE REJECTS AND INCIDENCES

Rejects and incidence reports may be resolved in several ways. They may be resolved by fixing the data during another extract from the source, fixing the data with the transformation program, or fixing the data in its loaded state. Another option would be to have procedures in place to correct the data after it is implemented into production, or have a procedure in place that provides a work around when the data is put into production. In some cases, all of the above is required to get the data to a production status.

PROJECT NOTES

Chapter 14

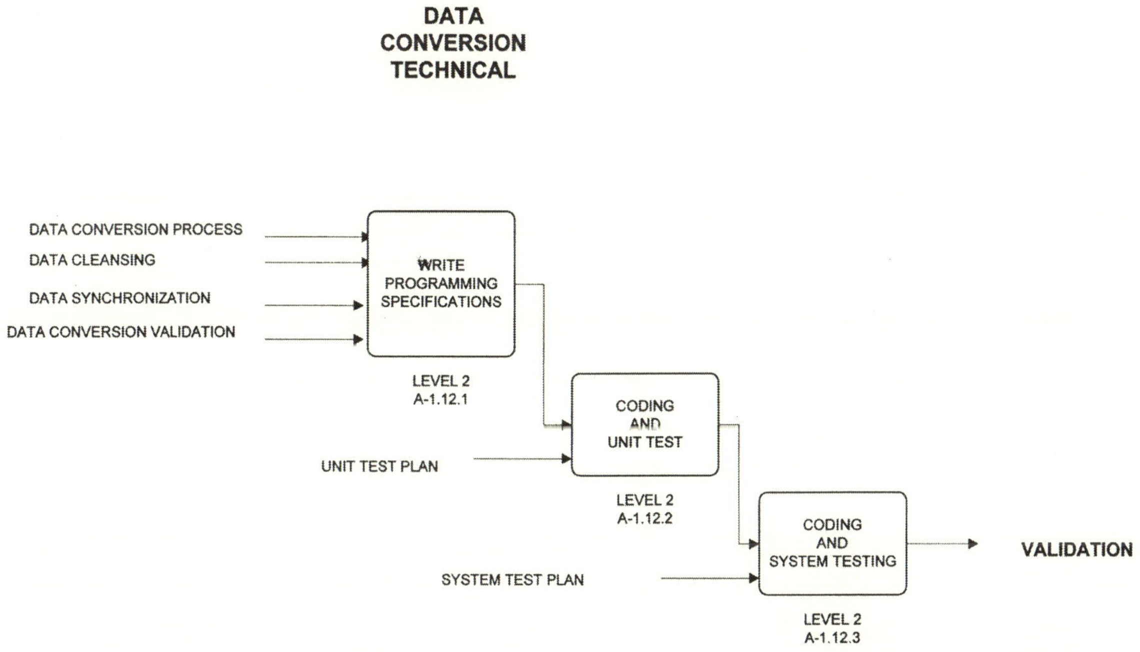

DIAGRAM 13 LEVEL 2 - Data Conversion Technical

PURPOSE:

The purpose of this section is to emphasize that a technical oriented group must be available to perform data extracts, data transformations, and data loads, and that there is a process involved with getting the programming and testing done.

LEVEL 2, NODE A-1.12.1 WRITE PROGRAMMING SPECIFICATIONS

All the user specification that are compiled as a results of the data conversion process, data cleansing, data synchronization, or data validation need to be turned into programming specification, or at least into a language the individual will understand that has to do the work to get the data extracted, transformed, and loaded.

LEVEL 2, NODE A-1.12.2 CODING AND UNIT TEST

Required programs are written and tested conducted as each unit of the programming work is completed. This process is vital in the conversion process and must be micro managed to assure that what appears correct on the surface for 25 sample records will not be a major problem when 1,000,000 records with all its variations of data elements are loaded.

LEVEL 2, NODE A-1.12.3 CODING AND SYSTEM TEST

As unit tests are completed, systems test, which includes two (2) or more components of the unit test must be performed. When this task is completed, data validation can occur.

Chapter 15

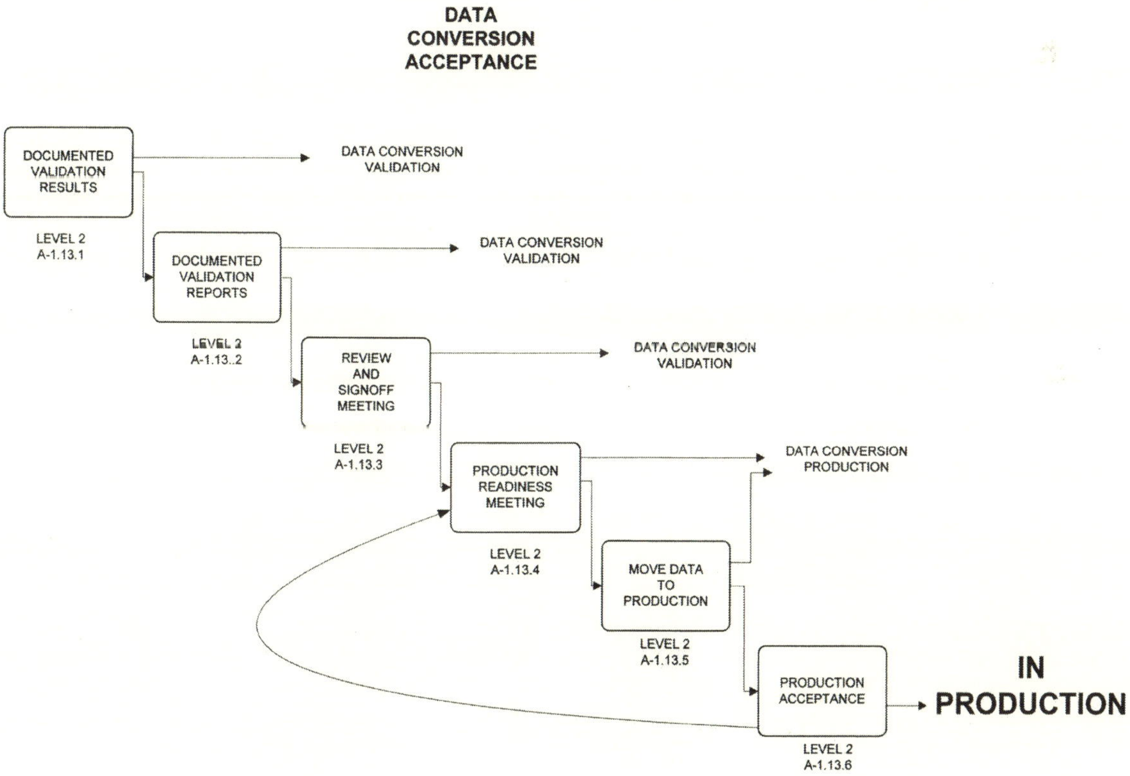

DIAGRAM 14 LEVEL 2 - Data Acceptance

PURPOSE:

The purpose of this section is to emphasize the steps required to move the data into production, after the data is converted and validated.

LEVEL 2, NODE A-1.13.1 DOCUMENTED VALIDATION RESULTS

The results of the validation execution must be prepared. The results could take the form of a reject report; the time the system test was performed, result of test based on incidence, and documentation of the AQL certification.

LEVEL 2, NODE A-1.13.2 DOCUMENTED VALIDATION REPORT

All validation results are combined into a validation report. The validation report should have an executive summary section where management can review summary information prior to sign-off. The summary information will list all the rejects by criticality, row counts, unexpected errors, and a recommendation from the conversion team as to how to proceed.

LEVEL 2, NODE A-1.13.3 REVIEW AND SIGNOFF

Prior to going into production, the data must be approved as production ready. The organization will probably require that the SME, Quality group member, and Data Owner all agree prior to going into production.

LEVEL 2, NODE A-1.13.4 PRODUCTION READINESS MEETING

All things must come together at the production readiness meeting. For the data conversion team, all validation reports must be signed off. In addition, the data conversion team should be prepared to present their "cut-over" plan.

LEVEL 2, NODE A-1.13.5 MOVE TO PRODUCTION

The moving of data to production must be carefully planned to cause the least amount of disruption to operations. Questions need to be answered in relationship to all elements involved in the moving of data. Will it happen at the end of the week, month, quarter, or year? How long will it take? What must be done prior to starting the moving of data? What support will be required?

LEVEL 2, NODE A-1.13.6 PRODUCTION ACCEPTANCE

After the data is moved into production, there should be a critical item validation test in production prior to acceptance.

PROJECT NOTES

Chapter 16

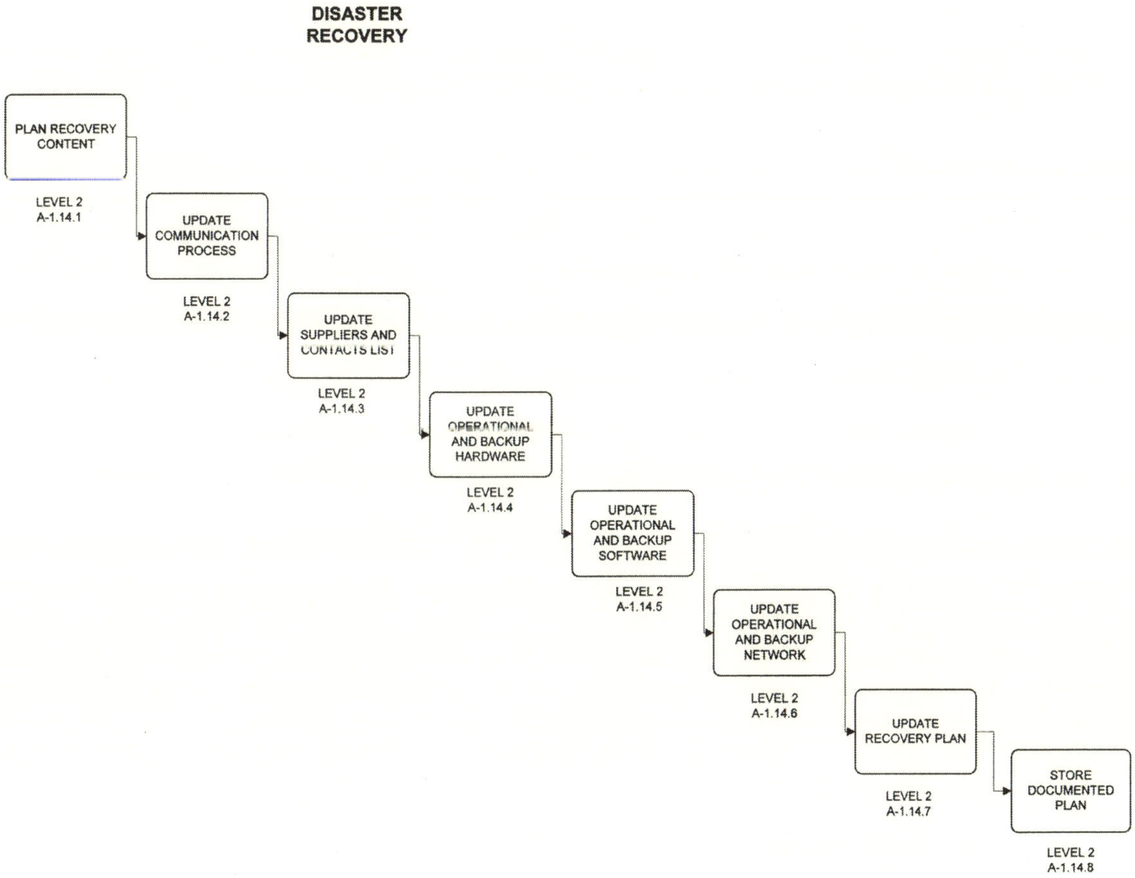

DIAGRAM 15 LEVEL 2 – Disaster Recovery Planning

PURPOSE:

The purpose of this section is to emphasize that the disaster recovery plan needs updating, or if one does not exist, to create one.

NOTE: If a new plan is to be created, then it is suggested that a consultant or organization that specializes in disaster recovery be hired to assist in the development of the plan. This flow diagram is to be used as a reference only and is not intended to furnish the details required to establish a disaster recovery plan or a contingency plan.

LEVEL 2, NODE A-1.14.1 PLAN RECOVERY CONTENT

Update the plan to reflect the critical business functions and the order of recovery. If it is a new plan, the items required under the content needs to be added. (Reference note in objective)

LEVEL 2, NODE A-1.14.2 UPDATE COMMUNICATION & PROCESS

If the new or updated software or hardware installed adds new functionality within the organization, new communications within the corporation need to be added to the communications plan, update accordingly.

LEVEL 2, NODE A-1.14.3 UPDATE SUPPLIERS & CONTACTS LIST

Evaluate the suppliers and contacts list. If new or updated software or hardware has been installed, then the contacts for the vendor and vendor list needs to be added to the plan.

LEVEL 2, NODE A-1.14.4 UPDATE OPERATIONAL AND BACKUP HARDWARE

If new hardware has been added to the facility, or if new applications are added to the hardware, then update both the operational hardware section as well as the backup plan for the hardware in case of an emergency.

LEVEL 2, NODE A-1.14.5 UPDATE OPERATIONAL AND BACKUP SOFTWARE

If new or amended software has been added to the facility, update both the operational software section as well as the backup plan for the software in case of an emergency.

LEVEL 2, NODE A-1.14.6 UPDATE OPERATIONAL AND BACKUP NETWORK

If the network has changed in anyway at the facility, update both the operational network section as well as the backup plan for the network in case of an emergency.

LEVEL 2, NODE A-1.14.7 UPDATE RECOVERY PLAN

With new equipment or software on the corporate premises, the recovery plan must be updated, considering all contingencies.

LEVEL 2, NODE A-1.14.8 STORE DOCUMENTED PLAN

As with the original plan, the document must be configuration controlled and stored according to the store requirements of the plan for the corporation.

PROJECT NOTES

Chapter 17

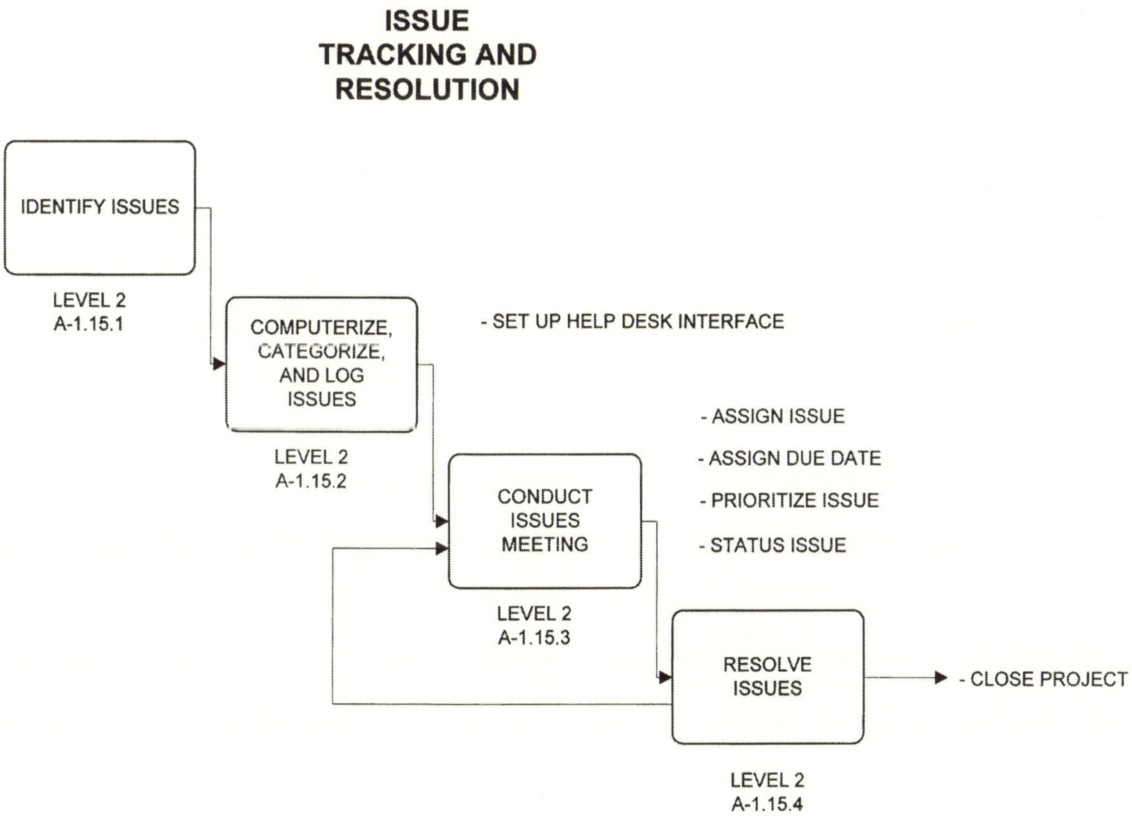

DIAGRAM 16 LEVEL 2 – Issue Tracking and Resolution

PURPOSE:

Issues that arise during the project must be resolve. Not only must the plan be in place as indicated in Chapter 6, but each issue recorded during the project must be resolves to the satisfaction of the user community and the project team.

LEVEL 2, NODE A-1.15.1 IDENTIFY ISSUES

Any problem or issue related to the success of the project must be identified. These issues may relate to the software, hardware, schedule, people, business process, data (both legacy and new), meeting rooms, dim lights, and a myriad of other issues that come up during the time line of a project. Some of these issues may seem minor or petty. However, each should be documented if for no other reason than to clarify the perceived problem.

LEVEL 2, NODE A-1.15.2 COMPUTERIZE, CATEGORIZE, AND LOG ISSUES

To assist in the timely documentation and resolution of issues, some sort of computerized program must be used. This computerized program could be a user written EXCEL program, an Access program, or a purchased issues log and resolution program. It should be recognized that the corporate Help Desk will generally have some sort of computerized program to assist the Help Desk in resolving problems. The program used by the project team should be different. It should be different because the project duration is short and the issues of the project should not be intermingled with those of the Help Desk. When problems arise on the project team that should be handled by the help desk, they should be noted by the project team and turned over to the help desk for resolution.

LEVEL 2, NODE A-1.15.3 CONDUCT ISSUES MEETING

All of the project team leaders must be invited to the issues meeting. It is in the meeting where issues are assigned to individuals, the due date assigned for each issue (and problems associated with the due date), noted as completed (if the problem is really resolved), and the category is changed if necessary. The meeting will be conducted by the Issues Lead person. Generally these meetings are conducted on a weekly basis in the early stages of the project. As the project end date approaches, the meetings are conducted more frequently until they are conducted daily at the critical stages of the project (at the end).

LEVEL 2, NODE A-1.15.4 RESOLVE ISSUES

All resolved issues are so noted and reported to the issues meeting. For those with a status of critical, actions should be taken by the Issues Lead person to follow-up on a daily basis to make sure some action is being taken to resolve the issue. These follow-up actions can also be reported during the issues meeting.

All issues that are not closed when the project ends should be handed over to the Corporate help desk for tracking. (Believe me, this happens.)

PROJECT NOTES

Chapter 18

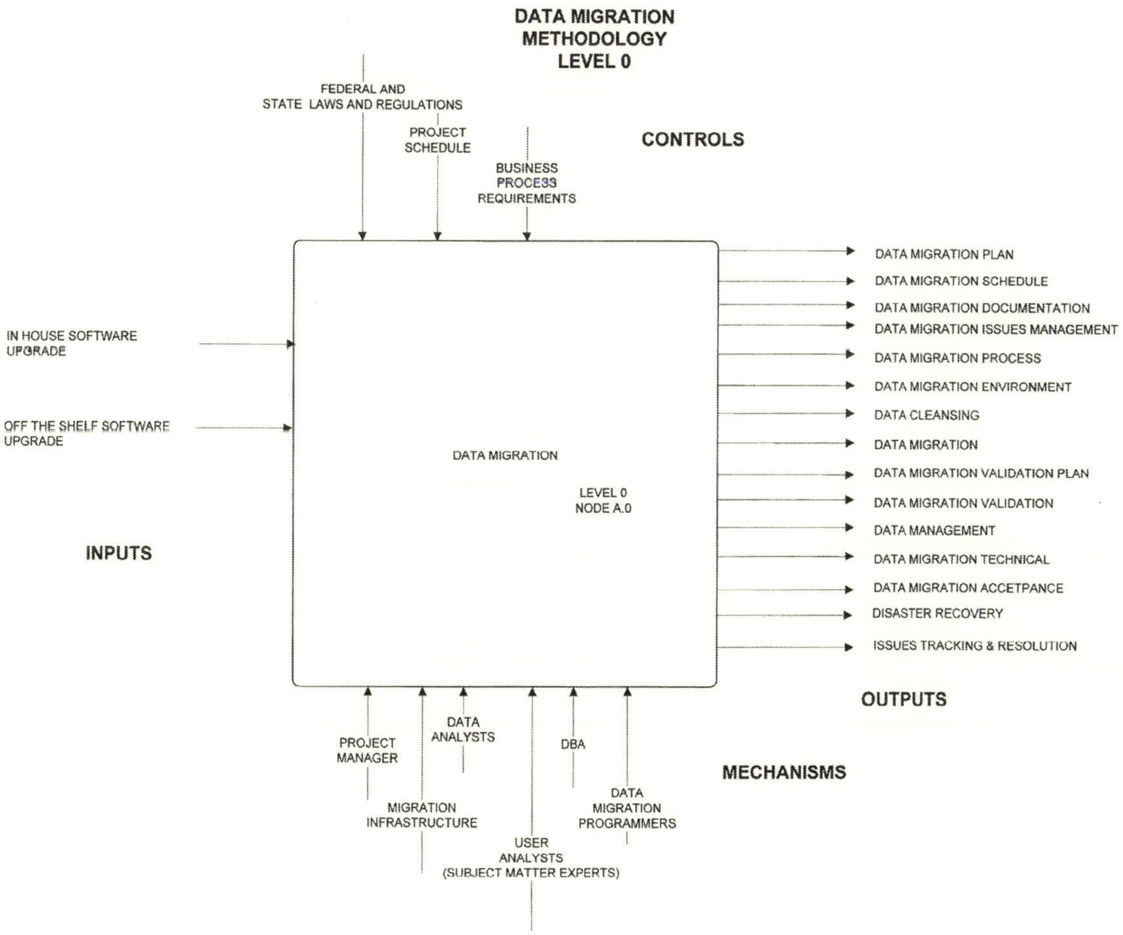

LEVEL 0, NODE B-0 Data Migration

PURPOSE:

The purpose of this section is to emphasize the importance of planning a data migration implementation plan, and executing to the plan. It is important to stress that the activities required for a data migration execution may be no less than those required for a data conversion project, and in some cases may be greater.

There is little to add in the explanation of the IDEF0 for data migrations that has not been recorded during the previous pages that refer to data conversion. However, some think that a migration may be easier than a conversion and do not spend as much planning time for a successful migration. This may not be the case and serious planning should occur, just as in data conversion. The levels of difficulty of a migration depend on several factors, or a combination of factors which will drive the project into disaster without proper execution of a well thought-out plan.

Factors contributing to levels of difficulty

1. **Configuration Control.** - If changes have been made to the software by the technical staff since the original installation, or if the data fields have been changed in house, the level of difficulty increases.
2. **Data Field Changes.** - If the software vendor has changed data field schema, definitions, or added new fields to the data base, the level of difficulty increases.
3. **Control Data.** - If the software vendor has changed configuration controls or data controls for the software, then these controls must be reviewed and changed with great care.
4. **Migration Leap.** - If the installation has not remained current with the recommended vendor upgrade schedule and the migration has to cover more than one migration upgrade level, then the level of difficulty increases.

5. **Reports.** - If the installation has produced many reports from the data tables, and if the data tables have changed for the migration, then the level of difficulty increases.
6. **Archiving.** - If the installation does not want to migrate all the data to the upgraded system but choose to archive historical data, then the level of difficulty increases.
7. **Personnel Resources.** - If the installation does not have anyone familiar with the changes that have been imposed on the new software, or changes made to the basic architecture of the applications, then the level of difficulty will increase.
8. **Hardware resources.** - If the migration includes migrating to a new level of the operating system, or new hardware whose operating characteristics are unknown, then the level of difficulty increases.
9. **Education and Training.** - If business processes are to change as a result of the migration, then the data migration becomes more urgent and stringent, hence more difficult.

These levels of difficulty point out the need for thorough planning for a data migration. Start at the top and follow through on all the IDEF0 diagrams.

PROJECT NOTES

APPENDIX

There are always helpful (hopefully) concepts available for data conversion. This appendix is provided to shed some light, and provide template samples of "what may be useful" data conversion concepts. It should be noted that some of these templates have been semi-automated for some of the vendors "off-the-shelf" software. While the automated software may serve to hasten the conversion/migration, the learning curve of the automation software, and the completeness of the mapping process may prompt one to perform mapping manually or in a less vigorous automated way.

Table Or Record Manual Data Mapping Template

TABLE OR RECORD DATA MAPPING TEMPLATE

The mapping view as outlined below is best suited to the technical oriented person that has to make sure the mapped data ends up in the right table.

NOTE: This template is to be used to map data in conjunction with the Data Conversion Process of the Data Conversion Methodology.

DATA TARGET TABLE OR DATABASE NAME	*TARGET FIELD NAME*	TARGET FIELD DATA TYPE	POSITION	REQUIRED	LEGACY FIELD	MAPPING RULE (Number Rules)
Target Table, Record, or Segment, to which data is being mapped.	*Data Dictionary name or number of target field to populate.*	*Specify type and length of target field (Character data, Numeric data, date, etc.)*	*Position of data field in Target record*	*Indicate if this field is a required field.*	*Data Dictionary name or number of Source field to be used to populate the target field*	*A1. Logic that will be applied to the source data to transform to target data requirements*
						A2.
						A3.

Notes:

1. Assign numbers (such as A1) to the mapping rules so that there will be a simple reference to the data element during the validation testing.
2. If the data is created for a field, provide the create algorithm (in the Mapping Rule column).
3. If special codes are used for a field, provide explanation of code.
4. If source definition does not match definition of target, provide explanation of reason for variant data.
5. If default data is used for a field, give brief explanation.

Panel And Table Manual Data Mapping Template

PANEL AND TABLE DATA MAPPING TEMPLATE

The mapping view as outlined below is best suited to the technical/user oriented person that has to make sure the mapped data ends up in the right table and on the right panel or screen.

NOTE: This template can be used to map data in conjunction with the Data Conversion Process of the Data Conversion Methodology.

DATA TARGET TABLE OR DATABASE NAME	*TARGET FIELD NAME*	**DATA TARGET PANEL NAME OR NUMBER**	*TARGET PANEL FIELD NAME*	**POSITION**	**REQUIRED**	**LEGACY FIELD**	**MAPPING RULE (Number Rules)**
Target Table, Record, or Segment, to which data is being mapped.	*Data Dictionary name or number of target field to populate.*	*Target Table, Record, or Segment, to which data is being mapped.*	*Data Dictionary name or number of target field to populate.*	*Position of data field in Target record*	*Indicate if this field is a required field.*	*Data Dictionary name or number of Source field to be used to populate the target field*	*A1. Logic that will be applied to the source data to transform to target data requirements*
							A2.
							A3.

Notes:

1 Assign numbers (such as A1) to the mapping rules so that there will be a simple reference to the data element during the validation testing.

2 If the data is created for a field, provide the create algorithm (in the Mapping Rule column).

3 If special codes are used for a field, provide explanation of code.

4 If source definition does not match definition of target, provide explanation of reason for variant data.

5 If default data is used for a field, give brief explanation.

Panel Or Screen Manual Data Mapping Template

PANEL OR SCREEN DATA MAPPING TEMPLATE

The mapping view as outlined below is best suited for the user person that has to make sure the mapped data ends up on the right panel or screen.

NOTE: This template can be used to map data in conjunction with the Data Conversion Process of the Data Conversion Methodology.

DATA TARGET PANEL NAME OR NUMBER	TARGET PANEL FIELD NAME	POSITION	REQUIRED	LEGACY FIELD	MAPPING RULE (Number Rules)
Target Panel or Screen Name or Number to which data will be displayed					
	Data Dictionary name or number of target on Panel or Screen.	*Position of data field on Panel or Screen (Header, Body, Footer)*	*Indicate if this field is a required field during the data loads*	*Data Dictionary name or number of Source field to be used to populate this target field*	*A1. Logic that will be applied to the source data to transform to target data requirements*
					A2.
					A3.

Notes:

1 Assign numbers (such as A1) to the mapping rules so that there will be a simple reference to the data element during the validation testing.

2 If the data is created for a field, provide the create algorithm (in the Mapping Rule column).

3 If special codes are used for a field, provide explanation of code.

4 If source definition does not match definition of target, provide explanation of reason for variant data.

5 If default data is used for a field, give brief explanation.

Report Request Template

REPORT REQUEST MAPPING TEMPLATE

NOTE: This template is to be used (generally by an end user) to request a report. It should be used prior to, and in conjunction with the Report Data Mapping template. Please modify to fit your environment.

	REPORT REQUEST
REPORT REQUESTOR	**Enter the name of the person requesting the report**
REQUESTOR DEPARTMENT	**Enter the department or organization of the person requesting the report**
REQUESTOR PHONE NUMBER	*Enter the telephone number or a contact number of the person requesting the report.*
REQUESTOR E-MAIL ADDRESS	**If e-mail is a standard way of communicating within your company, enter the e-mail address of the person requesting the report.**
REQUEST TYPE	**New report, modify existing, report, correct existing report (To modify or correct, only enter those fields below that are affected). Attach sample reports as necessary.**

	REPORT SPECIFICS					
PAGE NUMBER		Y				
RUN DATE		Y				
RUN TIME		Y				
SELECT DATA BASED ON			**1. XXXX 2. XXXX 3. XXXX**			
SORT REPORT BY:			**1. XXXX 2. XXXX**			
SEPARATE REPORT BASED ON (BREAK ON):			**1. XXXX 2. XXXX 3. XXXX**			
TOTAL ON:			**1.XXXX**	*Total based on breaks*		
REPORT TOTAL		**On last page**		*Yes*		
FREQUENCY	**On Demand, Weekly, Bi-Weekly, Monthly, Bi-Monthly, Yearly** Other					

REPORT HEADER						
HEADER FIELD NAME						
For each column on the report, enter desired column header names to go on the report if they differ from the panel name.	*Optional panel name will appear in the header columns if none is specified*					
REPORT FIELD DEF						
REPORT HEADER FIELD NAME	**PANEL NAME OR NUMBER**	**FIELD NAME ON PANEL**	**FIELD DESCRIPTION**	**SPECIAL RULE TO APPLY**		
Name of report field to populate. (Optional panel name).	*Specify panel name or number*	*Specify field name as it appears on the panel*	*Describe the field that you would like to see on the report.*	*Describe any special rule you would like to see applied to this field. If the rule applies to more than one field on the report, show rule at each field.*		

SAMPLE PROJECT SCHEDULE TASKS FOR THIS METHODOLOGY

TASK NO.	TASKS - LEVEL-0	TASKS - LEVEL-1	TASKS - LEVEL-2
1	**DATA CONVERSION PROJECT**		
2		**DATA CONVERSION PLAN**	
3			Data Conversion Scope
4			Data Conversion Topology
5			Conversion Data Types Or Application Area
6			Data Ownership
7			Environment Setup And Special Codes
8			Conversion Description
9			Conversion Testing Requirements
10			Change Process
11			Signoff
12		**DATA CONVERSION SCHEDULE**	
13			Develop And Refine Level One Tasks
14			Develop Level Two and Three Tasks
15			Establish Tasks Prerequisites
16			Document Assumptions
17			Resource The Schedule
18		**CONVERSION DOCUMENTATION MANAGEMENT**	
19			Document Available Requirements
20			Document Maintenance Requirements
21			Document Archival Requirements
22			Document Platform And Infrastructure
23			Document Communication Requirements
24			Implement Document Management System
25		**ISSUE MANAGEMENT PLAN**	
26			Document Issues Process
27			Document Tools Required
28			Determine And Document Issue Criticality Scale
29			Document Closeout Process
30		**DATA CONVERSION PROCESS**	
31			Identify Source
32			Validate Business Requirements
33			Identify Target
34			Plan Set Up Infrastructure
35			Evaluate Schema
36			Data Accuracy Assessment
37			Data Completeness Assessment
38			Defaults And Special Codes

39			Map Data
40			Define Variant Data
41			Mapping Descriptions And Transformation Rules
42			Mapping Specifications
43			Review And Signoff
44		**DATA CONVERSION ENVIRONMENT**	
45			Establish Region/Server Requirements
46			Establish Access, Security, Environment, And Codes
47			Determine Data Conversion Tools
48			Establish Region/Server Schedule
49			Establish Size Region/Server
50			Create Region/Server
51		**DATA CLEANSING**	
52			Review Legacy Data Accuracy
53			Review Legacy Data Completeness
54			Document Cleanup Requirements
55			Establish Clean-Up Infrastructure
56			Establish Clean-Up Process
57			Produce Data Clean-Up Specification
58			Review And Sign-Off
59		**DATA SYNCHRONIZATION**	
60			Analyze Legacy Functionality And Data Requirements
61			Analyze New System Functionality And Data Requirements
62			Determine Synchronization Infrastructure
63			Synchronize Data/Functionality Specification
64			Impact Analysis Report
65			Review And Sign-Off
66		**DATA SYSTEM MANAGEMENT**	
67			Business Requirements
68			Remaining Legacy Data Requirements
69			Archiving and Retirement
70			System Tuning
71			Operations Training
72			Operations System Set-Up
73		**DATA CONVERSION VALIDATION PLAN**	
74			Document Validation Process Flow
75			Determine Validation Requirements
76			Test Case Selection Criteria
77			Determine Criticality Scale
78			Determine Acceptable Quality Level (AQL)
79			Document Legacy Validation Base
80			Review And Sign-Off
81		**DATA CONVERSION VALIDATION**	
82			Execute Validation Plan (Selection Criteria)

83			Document Incidence
84			Execute Validation Plan (AQL)
85			Document Rejects
86			Resolve Rejects And Incidences
87		**DATA CONVERSION TECHNICAL**	
88			Write Programming Specifications
89			Code Programs
90			Unit Test Programs
91			System Test
92		**DATA ACCEPTANCE**	
93			Document/Organize Validation Results
94			Document/Organize Validation Reports
95			Review And Signoff Meeting
96			Production Readiness Meeting
97			Load Data Into Production
98 99		 DISASTER RECOVERY PLAN	Production Acceptance
100			Plan Recovery Content
101			Update Communication Process
102			Update Suppliers and Contact List
103			Update Operational and Backup Hardware
104			Update Operational and Backup Software
105			Update Operational and Backup Network
106			Update Recovery Plan
107			Store Documented Plan
108	DATA MIGRATION		
		Insert Level 1, 2, 3, ... as required using Data Conversion Tasks as a guide	

PROJECT NOTES

FOOTNOTES

[1] FIPS PUBLICATION 183

[2] In most cases, when implementing outside vendor "Off the Shelf" software, it is vital to the project to have one that is knowledgeable on the functionally of the product involved in the data conversion/migration project. Expected results from the execution of a transaction play an important role in the acceptance or rejection of the converted data. Therefore, knowledge of how the transaction should work cannot be stressed enough.

[3] Many of the application software vendors will have their own method of moving data from one location to another or transforming data. There are also vendors that market special software tools to access and view SQL or text data, move data across hardware platforms, view data on any hardware or software platform, and modify data. These tools generally are low cost and are easy to use.

[4] MIL-STD-105E

About the Author

Charles R. Scott has participated in over fifteen data conversion projects during the past twenty five years. He holds a BS degree in mathematics and a MS degree from the University of Southern California. His years of experience managing and participating in implementations of MRP, ERP, and EAP application Systems qualifies him in a unique way to develop and produce this methodology.

15585681R00074

Printed in Great Britain
by Amazon